Alewife

A Documentary History of the Alewife in Maine and Massachusetts.

Douglas H. Watts
with contributions by
Timothy A. Watts and Allan E. Watts

Foreword by Kerry Hardy

Poquanticut Press
Augusta, Maine and
North Easton, Massachusetts

Front cover photo: blueback herring and alewives at Presumpscot Falls, Falmouth, Maine, June 2009.

Back cover photo: Queequeg T. Dog, Ph.D. cleaning 100 years of garbage from China Lake Outlet stream, Winslow, Maine, after 2008 removal of Fort Halifax dam.

Published by
Poquanticut Press
131 Cony Street
Augusta, ME 04330
info@dougwatts.com

Foreword

by Kerry Hardy

I first met Doug Watts in 2007 at the Maine State House. We were both milling in the corridor, passing the time during one of those interminable waits that are part of the "public hearing" process. This is when citizens of Maine have the chance to tell various legislative committees what they think of the bills being considered in that particular session. At these same public hearings, oddly enough, there are decidedly *non-public* entities like corporate lobbyists and assorted agents of Maine's executive branch, and they too get to tell the legislators what's what.

As a result, about two out of every three speakers is an Augusta insider, dressed in a nice suit and getting a nice check to speak on behalf of the economic status quo. The odd man out is someone like Doug or me, who has to set aside their work for the day and travel to the capitol in hopes of making a difference for the lost causes of our day—things like education, health and welfare, environment, and other trifles that get in the way of business.

This setting was new to me, but Doug was already a seasoned pro; seasoned enough to know that we were engaged in the human equivalent of alewive behavior: beating our heads against a concrete barrier, in desperate hopes of obtaining what ought to be a simple birthright—the right to spawn and live, in their case; in ours, the right to have clean, functioning rivers.

On this particular day, it wasn't looking good for either the alewives or us. The legislators' brains were overcooked from considering the hundreds of bills that come forward each session, and the heavy hand of the state's executive branch was once again assuring them that all the necessary safeguards for anadromous fish were in place. As you'll learn from the essays that follow and the subsequent pieces of documentary history, the safeguards are indeed in place -- and there they sit, protecting our fish and rivers about as well as that rusted Revolutionary War musket nailed over the mantle protects your house.

By the end of that day, I was beginning to see what Doug already knew: when a system has been broken for two centuries, don't put your hopes on the ones who broke it being able to fix it. The challenge is to not so much to have *more* laws; it's to get enforcement of *any* laws. Laws are enforced by the executive branch of government, and since it's generally easier to buy one governor or president than hundreds of legislators, industry has chosen to heap the bulk of its blessings on the Blaine House throughout Maine's statehood. From their point of view, it's been a wise investment—hundreds of laws have been passed to protect anadromous fish, but *enforcement* of these laws has been virtually non-existent.

Doug was already well on his way to giving up on the legislature and taking the battle to the next level—the judicial arena. If the first two branches of government fail, perhaps the third still offers some hope? At this point in time, the answer is a qualified "yes"—but the real question is whether or not we, and the fish, will run out of time before we can legally compel the various executive agencies to enforce fishery laws. As George Mitchell said, it's a sad day when you have to sue the EPA to make them protect the environment; but like it or not, it's what we've come to.

For this reason alone, it's vitally important to have books like this one—books that *prove* that these fish were here, that we've passed law after law trying to protect them, that we've petitioned, pleaded, protested, and talked ourselves blue in the face trying to get enforcement, and that all the while we've seen the stocks dwindle from

4

the millions, to the thousands, to the hundreds, to the salmon run of four fish on the Kennebec that Doug tells of. Samuel de Champlain's words from 1605 *prove* that the St. Croix River had millions of alewives and stripers, and no corporate mouthpiece or DMR staffer can ever un-write those words. It's been said that Ed Muskie prevailed in his fight to get the Clean Water Act passed because he always had better information, and more of it, than his opponents did. As Doug would say it, these here are the *facks*—and without them, court battles cannot be won.

That said, there's an even better reason to have a book like this, and that's because there's still one more arena where battles await: the public mind. Good people will often, perhaps even *inevitably*, find their way around bad government, provided they have the necessary information, commitment, and passion to guide them-- and *Alewife* is a book that supplies all of these. Doug's book helps us *collectively* remember a fish we no longer *individually* know, and as we read his and Tim's essays from a lifetime spent wading through streams where fish are struggling for their lives, we cannot help but feel the urgency of the situation, and the sense of loss that looms for all of us.

In theory, the second half of this book *could* have been written by anyone with access to the various archives and historic texts—but in point of fact, it never *would* have been. But-- if you carry the childhood memories of silvery fish swarming up a brook, and carry the pain of seeing or collecting their poisoned, mutilated, or "dewatered" (such a lovely, guiltless way of saying "killed!) eggs and bodies from the streams that nature intended to be their homes—then, and only then, you might have the passion to write the first half of this book, and the drive to unearth the rich details of the second half, and then choose to share it all with the world. Thoreau asked rhetorically, "Who hears the fishes when they cry?" This is a book written by someone who does, and perhaps after reading it you will as well.

-- Kerry Hardy, Lincolnville, Maine.

Silly Verse

I saw a little brook
trickling through the woods
Its path was blocked with leaves
that fell from a nearby tree.
I took a stick and broke the dam
and the water shone up at me
and said, that is I think it said,
"Thanks for setting me free."

-- Allan E. Watts, October 1990.

An alewife's lot in life's to be eaten.
But this fate quite a few
Are good at defeating.
So they swim way upstream
All to fulfill their dream
To have kids in a cove
Where the sun gently heats them.

-- Douglas H. Watts, September 2010.

It's a Fack

Science books tell us the alewife (*Alosa pseudoharengus*) is a member of the herring family (*Clupea*) and is native to coastal watersheds of New England and the Atlantic seaboard of the United States and southern Canada. Like the Atlantic salmon, American shad, Atlantic sturgeon and striped bass, the alewife is a migratory fish species, born in freshwater and growing to adulthood in saltwater. Adult alewives reach a length of 14 inches and may live up to age 10. Alewives reach sexual maturity at age 3 or 4. Like Atlantic salmon, adult alewives use a finely tuned homing instinct to return from the ocean to the specific river system, tributary system and lake or pond where they were born.

In Maine, alewives swim up rivers to spawn in April and May. They spawn in freshwater ponds in June. In Massachusetts, and especially Cape Cod, alewives may return to spawn as early as March. Females broadcast their eggs into the water while males surround them and broadcast their sperm. After spawning, the adults swim back to the ocean. The eggs hatch in several days. Newly born alewives are transparent and one quarter inch long. They begin their life eating zooplankton, such as *Daphnia*, and aquatic insect larvae, and grow to two to three inches long in six weeks. These 'young of the year' alewives leave their birth ponds and swim to the ocean anywhere from late July to November, with most migrating in September and October at a length of four to five inches. Alewives spend three to four years in the open ocean, feeding on oceanic plankton until they swim back to their natal rivers and ponds to spawn. Alewives may spawn up to four times during their lives.

Alewives were found in every coastal river in New England. They were easy to catch in large numbers and could be smoked or pickled for year-round consumption and export. Alewives were particularly desired as fertilizer for corn cultivation and as bait for the coastal Atlantic cod fishery. Laws limiting the harvest of alewives were passed by many towns as early as 1700. In 1735, the Massachusetts Bay Colony passed the first of many laws requiring mill dam owners to provide passage for migrating alewives at their dams. Hundreds of laws were passed by the New England states to protect the alewife. Most of these laws proved ineffectual due to lack of enforcement.

By the early 20th century, the construction of mill dams without fishways had destroyed nearly all of New England's remaining alewife populations. In most watersheds in New England, alewives have now been absent for one or more centuries. Peoples' memory of their own backyard alewife runs, or even what an alewife is, has been almost totally lost.

As a kid the most scary episode of *Star Trek* was when Rojan, who looked like a game show host, hijacked the *U.S.S. Enterprise* to return to his home planet in the Andromeda Galaxy, 2 million light years away. To do so, he and his friends turned the entire crew of the Enterprise into baseball sized dodecahedrons of styrofoam which contained all of the chemicals and molecules which comprise living human body. If Rojan pressed a button on his belt loop, the piece of styrofoam would turn back into a living woman or man. If he crushed the piece of styrofoam in his hand, even his magic belt loop button could not bring the person back.

What Rojan did to humans is what we have done to the alewife. For centuries we have measured the value of this wild animal solely in terms of what its chemical constituents, as food for people, food for domestic animals, food for corn, can do for us; or whether we can make more money by fouling their rivers with dams and pollution and letting them go extinct in the process.

The documents in this book are a chronological record of observations made by many New England people over the past 400 years of the alewife runs in their local rivers and streams. They are presented *verbatim* to preserve their full context and original historic flavor, something that summary or paraphrase cannot do, and because many of them are not easy to locate in their original form in the various archives where they now reside.

What do Alewife think?

If a striped bass, eel or seal wrote this book it would read much like most of what has ever been written about alewives, ie. what has the alewife done for me lately? This is one reason a striped bass, seal or eel didn't write this book.

When I first tried to protect alewives in the early 1990s, the conventional wisdom of fisheries biologists was that alewives were 'weak swimmers' and could not ascend even small rapids or falls. Because these folks were all degreed fisheries scientists and I was not, I believed them. Until I spent a lot of time with alewives.

From 1998-2000 I traipsed a lot around Souadabscook Stream, a beautiful tidewater tributary of the Penobscot River in Hampden, Maine. The old falling down dam at its mouth had just been yanked out thanks to the Penobscot Indian Nation, the Atlantic Salmon Federation and John Jones, who owned it. Beneath the U.S. Route 1A bridge the Souadabscook goes over a natural falls that stairsteps down in four foot drops. At anything but full moon high tide the falls look like nothing but a fish with suction cups for fins could ever get over it.

I stopped by the bridge at Route 1A on a sunny May morning in 2000, as the tide fell and the ledges below the bridge got drier and steeper. A horde of alewives were gathering below the bridge like fans trying to get into a noon Red Sox-Yankees game. Because the water was so frothy, root beer colored and deep, it was hard to see anything, but in a few minutes I noticed there were a dozen alewives in the upper pools of the falls which had been empty a few minutes ago. I figured they must have been up there before I arrived.

The uppermost drop is the worst, an almost complete vertical drop of about 4 feet into a narrow bedrock plunge pool, and as the tide falls, its steepness increases. Now, I thought, how the hell are these alewives at the bottom of the falls going to swim straight up through this narrow, raging flume? Then I saw a few alewives scooting upriver at the very top of the falls, where it seemed impossible they could reach. I could not figure out how they were doing it. A transporter beam? Spontaneous generation?

So I leaned on my stomach over the edge of the ledge a few feet next to the torrent. I got a very brief glimpse of a 14 inch alewife frozen like the Statue of Liberty halfway up the falling water, her head and body pointed straight up, her tail beating 100 miles a minute but her body not moving up or down. In a blink she disappeared, which I assumed meant she fell back downstream into the plunge pool.

Then I saw another alewife, frozen like a still life painting halfway up the maelstrom, swimming as hard as she could but not making any headway, then she disappeared too. Nothing made sense.

Finally I crept really close to the falls, where the spray was totally soaking me, and adjusted my eye blink rate to about 1/30th of a second. I realized that every other time I saw an alewife fall back downstream after trying to swim vertically through the torrent, I was seeing an alewife applying a massive amount of burst swimming and squirting straight up in the air like a slippery watermelon seed and going *through* the falls.

My observation error was twofold: I doubted alewives had the

11

strength to swim vertically through the falls, and my eyes and optic nerve were too slow to see them do it. But as Henry Thoreau said, some circumstantial evidence is quite strong, as when you find a trout in the milk, or an alewife above an 'impassable' falls. With each passing minute there were more alewives above the falls at Souadabscook than below it, and there were more alewives at the foot of the falls waiting to take their turn. How can 30 million Elvis fans be wrong?

My brother Tim and I and his kids, Danny and Hallie, saw the same thing at the Nemasket River in Middleborough, Massachusetts a couple years later. The alewives in the Nemasket, facing a small vertical dam with a three foot straight drop, would attack it, swim straight up, suspend themselves vertically in the falling water, swimming as fast they possibly could and then disappear. Only after very careful observation, and watching a mother mink grab a few alewives above the dam did we all discover that a few milliseconds after they disappeared from sight we saw a small "V" wake above the upper lip of the dam. The alewives were getting over.

Since the Smelt Hill Dam at the head of tide on the Presumpscot River was removed in 2002, I have stood at the river's sharpest drop, a Class V rapids, just above the former dam site and wondered how alewives or shad or any fish can get over it. Like at Nemasket and Souadabscook I have watched the falls at the Presumpscot for hours and days, so close to its spray I was on the verge of falling in, and I still cannot figure how the alewives and shad get over it. But they do, since they can be seen upriver by the thousands now every spring.

From these observations I learned everything I had read about alewives being 'weak swimmers' was hokum. At Ticonic Falls in Waterville, Maine on the Kennebec River Tim and I have watched alewives negotiate ledge drops with just a half inch of water flowing over them. With an acute understanding of physics, the alewives turn on their sides and swim sideways up the ledges to maximize the surface area of their tails against the current and then right themselves as soon as the water gets deeper. When the current is hard they will swim so fast they shoot out of the water, not to jump, but because their momentum carries them out of the water and into the air.

Another thing I learned about alewives by watching them closely is when they are just babies, at the age of three months, before they go to sea for four years, they are happy, as in they like to play. As they move down river in the late summer and fall, the baby alewives will school up in small eddies in the river and jump clear out of the water *en masse* in tiny crescents, just like Atlantic salmon. They will only do this when there are no predators around.

In 2003, after the Edwards Dam had been removed from the Kennebec River in Augusta, I walked down to the dam site at sunset and sat on the bank next to the rusted steel, bricks, broken bottles and hobo shacks. Suddenly the entire river channel erupted into a bank to bank rain of baby alewives, each making little leaps above the water. It was if we had just been hit by a hail storm, but the sky was clear blue and the droplets were not hailstones. They were tens of thousands of three inch fish leaping.

A friend told me later, "They must have been jumping at a hatch," meaning hatching mayflies or caddisflies. But there were no hatching insects coming off the river. Me and the swallows, dragonflies and bats would have seen them. The baby alewives were jumping just for the sake of jumping – because apparently that's what baby alewives do on a late summer day at sunset. I knew what I had seen was something nobody had seen on the Kennebec for 200 years. It was cool.

When alewives are confronted with an outside threat, say an eel or striper or an otter or whale, they ball up. Each alewife wants to be at the center of the school, farthest from the threat. But every other alewife also wants to be at the center of the school, so the center of the ball constantly changes as the consensus of all the alewives as to what is the "real" center of school constantly changes. This reminds me much of humans' behavior when confronted by an outside threat. We all want to

13

be closest to the center of the crowd and farthest from the threat.

I believe fish think, although the concept is anathema to some human folk. Any good fisherman will tell you in detail how they think fish think, if only to outsmart them. Humans are the direct descendants of the direct descendants of fish. We are fish who lost our gills and traded fins for fingers. Sticklebacks built houses long before people existed.

When I watch a female Atlantic salmon who has not eaten a meal for six months spend six hours digging a nest in the riverbed and then stop, and suddenly decide to dig a brand new nest 20 yards upriver, I ascribe this behavior as thought, ie. the ability to make a considered choice, and realize over time you made a mistake and revise your initial choice. Either that or people in second marriages are salmon.

This not to say alewives and salmon are more like people than we think, but that we are more like alewives and salmon than we think.

Alewives and the Native People of New England

It is significant, to me at least, that when the old dam at the mouth of Souadabscook Stream in Hampden, Maine was removed in 1998, allowing alewives and salmon and eels and other fish to return to the river unimpeded for the first time since the 1700s, the event was sealed by a 'smudging' ceremony conducted by the elders of the Penobscot Indian Nation. The ceremony, which includes sweet grass being burned, is a ritual of sanctification and blessing.

Because this book relies upon preserved written records, it cannot describe 95 percent of the history of alewives and people, ie. the period from the first appearance of people in New England after the last Ice Age and prior to the first appearance of Europeans who left written observations which still survive. This period spans 8,600 years. Any volume which relies on less than ten percent of the total historic record is fatally flawed. That said, I believe over time we will be able to greatly augment our knowledge of this 8,600 year period if we diligently endeavor to do so.

Humans most recently began living in New England about 9,000 years before present, when the region was still thawing out from the last Ice Age. New England was a much colder and drier place then, with a landscape and climate more resembling the taiga and tundra of northern Canada.

It is uncertain when alewives first began coursing their way up coastal New England rivers to spawn in ponds, however we can bracket the date within a few millennia. A wooden stick fish weir discovered in the 1980s at the inlet of Sebasticook Lake in central Maine has been accurately dated using radiocarbon techniques. This dating method shows the Sebasticook Lake fish weir was in operation at least 6,000 years before present, ie. 4,000 B.C. For comparison, the Great Pyramids at Giza, Egypt were built in 2560 B.C.

The weir's location, at the inlet rather than the outlet of Sebasticook Lake, suggests it was used to capture American eels migrating downstream from Wassookeag Lake in Dexter to the ocean. The presence of American eels at this great distance (100 miles) from the Atlantic Ocean more than 6,000 years ago suggests other migrating fish species, such as the alewife, were present in coastal watersheds at this time.

Archaeological excavations across New England in recent years have revealed preserved bones of alewives within the cooking hearths of Native villages and encampments. These sites range back to 4-5,000 years before present. Because the acidic soil of New England tends to destroy fish and animal bones, the presence of alewife bones at these and other prehistoric sites is undoubtedly under-represented.

In recent years, the authors and others have located a number of stone fish weirs in various rivers in New England, most previously unknown to archaeologists and historians. Stone weirs cannot be directly dated using radiocarbon techniques since carbon from once-living tissue is required. However, prehistoric stone artifacts at these sites provide a window for when these weirs were used.

The age of datable artifacts at these weir sites compare favorably with the 6,000 year age documented for the Sebasticook Lake wooden fish weir. The location of these weirs, all directly below lakes known from historic records to have supported very large alewife populations; and their orientation in the river bed in an inverted "V" or "W" with the points facing upstream, shows they were designed to catch fish migrating upstream. Based upon modern scientific understanding of the productivity of alewives in relation to the surface acreage of their spawning ponds and lakes in coastal river systems, each year the prehistoric weirs at these sites would have intercepted alewife runs of anywhere from 250,000 to 1-3 million adults. At these sites the alewife was by far the most numerous fish.

Perhaps the best picture of how the Native people of New England used and celebrated the alewife for millennia was found several years ago by my brother, Tim Watts, in the musty bowels of the Taunton Historical Society collection in Taunton, Mass. This description, which dates to the 1600s, describes Contact Era gatherings of Pokanoket Wampanoags at the mouth of Cohannet, now called Mill River, where it enters the Cohtuhticut, or Titicut, or Taunton River, in what is now downtown Taunton, Mass:

"The ancient standers remember that hundreds of Indians would come from Mount Hope and other places every year in April, with great dancings and shoutings to catch fish at Cohannit and set up theyr tents about that place until the season for catching alewives was past and would load their backs with burdens of fish & load ye canoes to carry home for their supply for the rest of the year and a great part of the support of ye natives was from the alewives."

It seems likely these spring gatherings of native people were conducted throughout New England, wherever the alewives ran in great number, for thousands of years before any writer was around to describe them.

What's in a Name: Madamas, Amasa, Namas, Herrin, Alewife.

The native languages of New England are rife with words and place names referencing the alewife. Since these languages are all oral, spellings are phonetic approximations of the sound and syllable as spoken. The Penobscot language gives alewives the name "madamas," with a variety of spellings depending on the writer. Sites where large numbers of alewives were caught carry the name "Madamiscontis" or "Mattamiscontis," the prefix referring to alewives, the suffix referring to a place where they could be caught in abundance. Two streams flowing into the Penobscot River still carry these names: Mattamiscontis Stream at the lower end of the Piscataquis River near Howland, Maine and Blackman Stream, aka Madamiscontis, in Bradley, Maine.

To the west, in the heart of the Kennebec Indian territory, Farmington Falls on the Sandy River has the very old name Amasacontee, referring to a place on the river, a village and the tribal name of the people who lived there in the 1600s and undoubtedly much earlier. The prefix 'Amasa' is similar to the Penobscot "madamas" (alewife) and the suffix, 'contee' is the Kennebec version of the Penobscot 'contis' for fish catching place.

In southeastern Massachusetts, the word for alewife is yet another variant of that used by the native people of Maine. The river which forms the outlet of Assawompsett Pond in Middleborough and Lakeville is called the Nemasket. The Assawompsett Pond complex, which includes five large interconnected natural lakes and is the largest group of natural lakes in Massachusetts, is home to the largest alewife run in New England, numbering from 1-2 million adults annually. The prefix for Nemasket combines the Wampanoag word for alewife (Namas or Nemah) with a suffix (ket or kitt), denoting a place where fish are caught in abundance. A 1668 record from Plymouth gives the spelling of Nemasket as "Namasakett."

So over the span of 250 miles of New England, from the mountains of western Maine to the boggy spruce woods of north central Maine and down to the comparatively balmy, holly treed climes of southeastern Massachusetts, we have the same word and meaning filtered through local dialects, Mattamiscontis, Amasa-contee, Namassakett: all meaning a place where lots of alewives can be caught.

This suggests it is very doubtful the word 'alewife' came from the languages of the Native people of New England. In maritime Canada, due to the strong French influence there, alewives are called 'gaspereau,' a word which like 'alewife' has no obvious connection to any native languages. Alewives are not native to Europe; they are only found along the eastern North American seaboard from the Saint Lawrence to Florida. However, alewives bear a close resemblance to the ocean herring (*Clupea harengus*) which is native to the ocean waters off the U.S., Canada and northern Europe, thereby explaining their scientific name, *Alosa pseudoharengus*, meaning literally a "pseudo" or false ocean herring.

Two of the earliest European references to alewives, by John Pory in Plymouth, Mass. in 1622 and John Josselyn in Scarborough, Maine in 1674, use the word alewife, or in Pory's case 'old wife,' and the word 'herrin,' noticeably without the 'g' at the end. It is presumed the word 'herrin' was used by early European settlers because of the strong

19

resemblance between the alewife of the New World and the oceanic herring known in Europe; and that Pory and Josselyn learned the words 'herrin' and 'alewife' or 'old wife' from European settlers in Mass. and Maine, rather than from Native informants, since neither author references the native words which by this time had long been in use.

Josselyn noted, "The Alewife is like a herrin, but has a bigger bellie therefore called an Alewife." Anatomically this observation is true, since the alewife has a much deeper body (from back to belly) than the ocean herring. While speculative, it is at least plausible that the term 'alewife' or 'old wife' was a sexist play on the deeper or bigger belly of the alewife as compared to ocean herring with the appearance of an older English woman of somewhat portly dimensions.

Longtime residents of southeastern Massachusetts do not call alewives alewives. As in 1622, they are still called 'herring' or 'herrin.' If you call alewives by the name 'alewife' on Cape Cod or Buzzards Bay you will often draw a blank stare unless you say what you *really* mean is 'herring.'

This is why the towns of southeastern Mass. and Cape Cod have had for decades and centuries "Herring Commissions" and "Herring Committees" appointed at town meeting to supervise and oversee the town's "herring runs;" and one of the largest alewife spawning ponds in Massachusetts, at Manomet near Plymouth is called on maps "Great Herring Pond" and the outlet of the pond at the Cape Cod Canal in Bournedale is known as the "Herring Run," as is the motel where the fishway and outlet are now located. A dozen miles to the west the spot on Route 6 where it crosses the Mattapoisett River at the Mattapoisett/Fairhaven line has long been called the "Herring Run." Various historic weirs along the Mattapoisett River for catching alewives have long been called "Herring Weir" (Upper or Lower) or similar names. None of which use the word 'alewife.'

Even as a kids in the early 1970s, when our mom and dad brought us to the herring run each spring to catch them with our hands, usually in Pembroke or Middleborough, neither our parents or anyone else called them 'alewives.' They were always called "herring"; and the place you went in April to watch and catch them was "the herring run."

Whether you call them madamas, amasa, namas, old wife, alewife or herrin, they are a remarkable native wildlife species who still survive despite all our misguided efforts in the past 400 years to drive them to extinction. They will come back again in enormous numbers in just a few years if we get out of their way.

Why do Alewife
go out to the sea and back?

Trying to be helpful, a human guest might ask an alewife, "Why not just choose between fresh and saltwater? That seems to be the source of your problem." To which the alewife might ask, "Why don't you choose between a house and a job?"

Prior to the early 1800s, most of the fish species in the fresh waters of coastal New England spent much of their lives in saltwater. The habit of being born in freshwater but living to adulthood in saltwater is called anadromy. These fish species are called anadromous. [1]

---

1 Aside from various small minnows and darter species, the number of non-anadromous fish species native to New England is sparse: white sucker, chain pickerel, pumpkinseed, hornpout, brook trout, lake trout, fall fish, yellow perch. New England's tribe of sea-run fish includes Atlantic salmon, American shad, Atlantic sturgeon, shortnosed sturgeon, alewife, blueback herring, striped bass, white perch, rainbow smelt, American eel, sea lamprey, sticklebacks and tomcod. Chain pickerel and yellow perch were absent from much of northern New England until being introduced by people in the early 1800s. This disparity may be explained by the extremely harsh conditions placed upon fish living solely in freshwater in New England during the 100,000 years of Pleistocene glaciation, which most recently ended 12,000 years ago, and the comparatively more hospitable conditions for fish who can live in salt and fresh water.

There is good evidence salmon evolved as freshwater fish and only moved into saltwater as adults. The origin of alewives and their close cousins, the blueback and American shad are more equivocal.

Salmon are closely related to freshwater fish such as brook trout and brown trout and have no close relatives which live entirely in the sea. In the opposite case, alewives have no close relatives in freshwater, but do have a close relative, the Atlantic herring (*Clupea harengus*), which lives solely in saltwater.

The fertilized eggs of all three of the river herring (alewife, American shad and blueback herring) require fresh water or water with low salinity to hatch. Of the three, alewives are the most tolerant to salinity when hatching. This tolerance is not total, however, since there are no alewives known which can live in saltwater alone. If this were true, none of the hundreds of alewife runs along New England's coast would have died out due to impassable dams, which they did, since the alewives could have just spawned in saltwater along the coast.

In certain conditions alewives can be induced to live out their full life in large freshwater lakes and artificial reservoirs made by dams. While humans can deliberately induce alewives to live in lakes and ponds without ever going to sea, there is no evidence of any natural populations of freshwater alewives anywhere in their range. All native, naturally occurring populations alewives spend most of their lives in the sea and return to fresh or brackish water to give birth.

The origin of anadromy is unclear. Of all the thousands of fish species in the world, most are strictly freshwater or saltwater. Put a bluefish in freshwater and it dies. Put a smallmouth bass in saltwater and it dies. This is for the same reason that we cannot drink saltwater. In freshwater, fish have a shortage of salts and must retain them. In saltwater they have too much salts and must expel them.

Anadromous fish species are peculiarly common in northern latitudes, become increasingly scarce as you go south and practically

23

disappear at the Equator. This suggests there is an advantage for fish in northern climates to spend more time in the ocean rather than in freshwater; and this advantage appears to diminish as you move south.

Anadromy places two severe penalties on a fish. First, the fish must devote scarce resources to the very complex physiological mechanisms necessary to live in both fresh and saltwater for various parts of its life. Second, to travel between fresh and saltwater the fish has to migrate, often for very long distances. It seems to make no sense for anadromous fish to have evolved in the first place. Why not live entirely in fresh or saltwater? Since anadromous fish do exist in great numbers there must be some 'buried' benefit to this behavior which outweighs its costs.

As noted by fisheries scientist Jim Lichatowich in *Salmon Without Rivers*, the reason for the ubiquity of anadromy in the north and its absence toward the Equator seems to be that in the north, the ocean is more productive than its adjacent freshwaters, while near the Equator the freshwaters are more productive than the adjacent ocean.

Speaking of Pacific salmon, but in an argument equally applying to alewives, Lichatowich writes: "Salmon grew rapidly in the oceans, and their larger size at maturity improved their fitness in the rivers of the Northwest ... Anadromous salmon could produce more offspring because the larger females carried more eggs to the spawning grounds. Furthermore, the larger, stronger fish were more able to muscle their way over falls or through strong currents, extending their distribution throughout the Northwest's network of rivers ... Apparently the Pacific salmon abandoned freshwater to rear in the more nourishing oceanic pastures of the northern latitudes."

A benefit of anadromy not mentioned by Lichatowich is that anadromous fish are not in a position to inadvertently eat their own young. Most anadromous fish, including alewives, cease to eat during their spawning migration in freshwater and instead rely on stored muscle mass. And once they have spawned, they quickly return to the ocean and away from where their babies are hatching and growing.

In Atlantic salmon, this situation is extreme, with adult salmon not eating for their entire sojourn in freshwater and you can see why. Atlantic salmon typically reach their spawning grounds 3-4 months before they spawn. Most of the small fish in the spawning grounds are baby salmon. If the adult salmon's appetite was not physiologically repressed once it enters freshwater, they would end up eating most of the baby salmon in the river. Not eating for months on end places great stress on adult salmon, shad and alewives; many do not survive the rigors of spawning to return to the ocean. But apparently the trade-off works. From a salmon or alewives' balance sheet it makes more sense to risk dying from lack of food during spawning than resorting to eating your own young.

There are populations of alewives in the United States which never leave freshwater. They are called 'landlocked' alewives, rarely get more than four to five inches long and live out their lives in the lakes and ponds where they were born. There are no natural populations of landlocked alewives; all known populations were created by the deliberate or inadvertent actions of people.

The alewives in a landlocked population are basically dwarves. During their whole lifetime they never get bigger than a sea-run alewife gets during its first months of life. This is not surprising since female alewives produce enormous amounts of babies, alewives eat a lot, and there is a lot less food in a lake than the Atlantic Ocean.

In their sea-run lifestyle, alewife mortality is very high, so females compensate by producing lots and lots of babies. In a 'landlocked' population the fish never leave the lake so mortality due to migrating and predation is much lower. Without this mortality pressure, the alewives are free to eat most of the zooplankton in the lake, and when they've eaten nearly all of it their growth rate slides to zero, but enough make it to spawning age, due to their prodigious fecundity, to continue the population.

Against some very stiff competition, like poisoning streams with rotenone to count how many fish are in them, perhaps the dumbest idea ever devised by fisheries biologists was putting sea-run alewives in large lakes remote from the ocean where alewives never naturally existed. Once you understand anadromy you can see why. Sea-run female alewives must produce a gazillion babies because so few will ever get back to the ocean, grow for three years, and successfully migrate upstream to spawn themselves. The babies, born in early summer in freshwater spawning ponds, must eat voraciously during a short summer growing season to get big enough to make it back down to the ocean.

If you put alewives into interior lakes where they are not native, but are big enough so they can reach spawning age without ever leaving the lake, all hell breaks loose. The alewives, released from the extreme mortality pressure of their natural, sea-run lifestyle, run amok and consume most of the zooplankton in the lake, and when that is gone, start preying on the young of the fish native to the lake, including predator fish, eliminating the fish which can grow big enough to eat them.

For these reasons, much misguided antipathy has developed around the alewife as a species of native wildlife. But who caused the problem? We did. We deliberately dumped buckets of them in places where they never lived in the first place. Or, in the case of the Great Lakes, we built a shipping canal around Niagara Falls which allowed them to enter the Great Lakes, where they never existed. Alewives can't build shipping canals around giant impassable waterfalls or put themselves into buckets and trucks and travel for hundreds of miles to a remote, inland reservoir. But we can, and we did.

In New England, and especially in Maine, tales of these disastrous man-made introductions of alewives into non-native waters has poisoned the well of efforts to restore native, sea-run alewives to the streams and ponds where they have lived since the last Ice Age. It's easy to push a button on your computer and print out a newspaper article or paper about the effects of the manmade introduction   to the Lake

Michigan, where they never lived until we put them there. Apparently, it's not as easy to look up in a dictionary the words "native" and "non-native" and note the distinction.

The entire 200 year history of our efforts to 'improve' and 'augment' the fish life in our rivers and lakes by putting into them fish which never lived in them reminds me of a joke by my wife Lori, who said while doing carpentry, "No matter how many times I cut this, it's still too short."

A few years back, I sat at a meeting about taking the first dam out on Cobbosseecontee Stream, just down the street, to allow native sea-run fish to get back up to their native habitat in Cobbosseecontee Lake for the first time in 240 years. Bill Monagle, a nice guy and skilled water quality scientist employed by the Cobbossee Watershed District, said he was dead set against letting any native alewives return to the lake, for fear they might harm the lake's declining water quality. Coining a new term, he called alewives 'native invasives.' Like a great horned owl, my head did a near 360. How can a native species be an invasive species? These are antonyms. Calling a wild animal a 'native invasive' is like calling a house paint 'black-white.'

While I disagreed with Bill's conclusion, I knew where it was coming from. During their two months between hatching and going out to sea, baby alewives live on zooplankton, the tiny little critters which eat phytoplankton, algae. Algae is one of the most undesirable and most visible end products of humans putting too much phosphorus into lakes and ponds, via septic systems, land clearing and lawn fertilizers. Too much algae makes once crystal clear ponds turn a murky brownish green. Zooplankton, like Daphnia, happily munch on phytoplankton all day long and keep the lake from getting as brownish green. Baby alewives (and all baby fish, as well as dragonfly and damselfly larvae) happily munch on Daphnia. So Bill figured the last thing he needed, since his job was trying to keep the lake from going completely murky green, was a few hundred million baby alewives eating all his prized Daphnia and thereby letting the algae run amok.

I admired and respected Bill's Sisyphean effort to keep Cobbosseecontee Lake from sliding down the slope from crystal clear to putrid green brownishness, as so many other Maine lakes have descended. I certainly didn't want to do anything that might increase the slope of the slide. Only later I realized Bill and I were looking through different lenses at the very same star. It all revolved around how you define 'health of a lake' and its antonym.

Bill viewed the health of Cobbosseecontee Lake primarily through the lens of water clarity, chlorophyll concentrations, and total phosphorus in the water column. I viewed the lake's health primarily through whether its keystone native species, particularly alewives, actually lived in it. Bill's job, as he saw it, was to do whatever he could to reduce phytoplankton growth, or in the alternative, encourage lots of zooplankton to eat the phytoplankton. My job, as Bill saw it, was to restore a native fish species to Cobbossee Lake, the native baby alewife, which is one of the most efficient predators of zooplankton on Earth.

But I knew, and Bill knew, that long before Boston real estate speculators 'discovered' Cobbosseecontee Lake in the late 1700s, the lake was filled with millions of baby alewives each summer and the lake was crystal clear and this balance had existed for millennia. In the fall, the baby alewives swam down out of the lake to the Kennebec River and the Atlantic Ocean. What had changed was us. Since the 1700s it has been our influence on Cobbosseecontee Lake that has caused it to veer toward a murky green. It hasn't been alewives, since they have not been allowed to swim into the lake and spawn for the past 240 years. How can alewives be blamed for something they never did, and before they even have a chance to do it? My brother would say it's a 'fish iniquity.'

After our meeting, it occurred to me the ideal solution to Bill's dilemma and job description would be to chlorinate Cobbosseecontee Lake like a giant swimming pool, thus keeping any phytoplankton at all from growing. With enough chlorine, on a clear day you could see the stones at the bottom of Cobbossee Lake in 60 feet of water. But that would also kill everything in and along the lake. So I thought, Bill's real goal is not making the lake as clear as possible even if it means sterilizing the lake. So that means Bill considers the critters in the lake to be important. In fact, the whole enterprise is about keeping the lake healthy for the sake of all the critters in the lake. But which critters? By what standards and parameters? Are some allowed and some denied? Who gets to play God?

Osprey, loons, ducks and eagles don't need to ask my or Bill's permission to live along Cobbosseecontee Lake. They fly there. Nor do squirrel or deer or moose or muskrat. All the studies in the world will not stop a bald eagle from building its nest in a tall pine next to Cobbossecontee and reclaiming its native ground whenever it wants to.

But migrating fish, like native alewife, must request our permission do what the bald eagle does because the only route to their ancient homes is through narrow, windy rivers which we decided many decades or centuries ago to block with stone and concrete dams.

I will always accept the sincerity of Bill Monagle's desire to save Cobbosseecontee Lake. But I will always have to ask: What are you saving it from? Whom are you saving it for?

Nemasket:
The River of Smiles

By Timothy A. Watts

In the Wampanoag language Nemasket means place of fish. Today as in days of old this name suits the Nemasket River well. Each spring as early as February they begin arriving, at first alone as scouts and then in small groups. As the spring sun rises higher and the water warms they swarm up Nemasket by the tens of thousands. By late April and early May the fish flowing up stream seem to overwhelm the water flowing down. At the end of the run in early June more than one million of these fish will have made their annual journey up the Nemasket.

It is here at Nemasket, perhaps more so than anywhere else in New England, that ancient cultures of the past join hands with our modern one. Like an unbroken common thread, Nemasket flows through us to connect the Ancient Archaic to the later Woodlands Period and on to our modern culture. For near ten thousand years people have continuously come to the riffles at Nemasket each spring to greet the return of these fish called Alewives.

In times past the reasons for coming here were clear: food, irreplaceable sustenance for both people and their crops. Today the reasons are not so clear. Although some folks still use the fish for a fertilizer and others still fry their roe to eat and still others catch them to bait larger fish, there is something else that brings us to this place.

What is it about Nemasket and these fish that draw us here each spring? One common theme is, "we simply like them." Naturally, the follow up question would be why do you like them? More often than not people respond to this question with a simple shrug of the shoulders and a smile. For many people including myself, it is the fond childhood memories of warm spring afternoons spent scooping alewives from the water with bare hands. It is also the spectacle of seeing so much life rippling through such a narrow space. There is also the "underdog" factor, where we instinctively feel for a creature who against long odds struggles to reach its birthplace to spawn a new generation.

The Nemasket River maintains the largest run of alewives in New England. This is in large part due to the wide pristine waters of the Assawompsett Pond complex in Middleboro, Lakeville and Rochester. The outlet of Assawompsett is the beginning of the Nemasket River. Dr. Maurice Robbins in his book <u>Wapanucket</u> states that "In pre-colonial times the Nemasket River flowed out of the lake at a point some distance east of its present location. An earthen dam now crosses the ancient bed and parallels the shore of the lake."

Apparently at some point in colonial times they moved the outlet of Nemasket to its present location. This is visible when approaching Assawompsett by canoe. About a hundred yards from the outlet the river widens and its course runs almost perfectly straight toward the pond. Apparently some enterprising souls attempted to channel the Nemasket for a shipping canal. They either ran low on shovels or strong backs; fortunately for us and the Nemasket the scheme was a failure.

From its outlet the Nemasket meanders lazily through marsh and swamp lands until it goes under Route 495 and then Route 28. Passing beneath Route 28 heading downstream, the new Middleboro Little League field would be on your left. During construction of the field they unearthed an Ancient Wampanoag village. Unfortunately they dug up and hauled most of the site away before it could be well documented. The remaining artifacts suggested that the site was several thousand years old, and was probably a heavily used area in ancient times.

Below this point the river continues down to the dam, and the alewife fishing site below it at Wareham St. The Nemasket scrambles down one of its few riffle reaches here, leading to a short stretch that brings you to the Ancient Wading Place at the Route 105 bridge below the center of town. Traveling further down through more meadow and swamp lands, you come to the place called Muttock, otherwise known as Oliver's Mills, at Route 44. This was the site of another extensive Wampanoag village and fishing site which was used from ancient times to the colonial period. Where the bones of the old mill complex now litter the river there was once a stone fish weir used to catch alewives and shad. The Wampanoag village and ancient burial place sat above on the hills over looking the stream to the south and east.

Once past here the Nemasket continues its meandering course down under Rt. 44 past the Middleboro Sewer Plant and on into the peaceful undeveloped marshes of North Middleboro. It is about half a day paddle from here to the Nemasket's confluence with the Taunton (a.k.a. the "Great River") below Titicut St. in Bridgewater.

One other suitable name for the Nemasket might be the river of smiles. As a resident of Middleboro I have the privilege of being a voluntary observer for the Middleboro Lakeville Herring Fishery Commission. Each Sunday morning during the fish season I go down to the run to check permits and keep an eye on the goings on.

People come from all over to see the spectacle of the Nemasket Run. Adults and children scoop the fish up, dumping them into buckets to take home. Children scamper around, trying to pick up the fish that flop out of the buckets. Across the stream a mother mink darts down to the water to snare an alewife from the shallows; people pause their fishing for a moment to watch her haul it back to her den. Down below the fishway, soaking wet kids thrash about in the shallows like a gaggle of bear cubs on a salmon stream. Oblivious to the cold they scoop the fish onto the muddy bank with their hands and wrestle with each other for the silver trophies fresh from the sea.

One particular afternoon I happened to stop by the run in the early afternoon. Teachers from the local school were just arriving with a group of "special needs" kids. It was a perfect afternoon for catching, warm and sunny, and the river loaded with fish. Teachers and chaperones wheeled the kids in their wheelchairs down to the river bank with nets in hand. It was a sight that could bring tears to the eyes of anyone with even half a heart.

I never saw a group of kids have so much fun. The teachers and chaperones had all they could do to keep them from plunging into the water. One would brace the chair, while another would hold the kids by the shoulders as they lunged out with their nets. Then another would have to help them haul up their heavy loads of fish and release them, only to repeat the seen all over again. When it was time to leave, all were tired, thoroughly soaked, covered with fish scales, smelly and grinning from ear to ear.

On another morning I was doing my watch at the run when a very old woman arrived with what appeared to be her granddaughter. It was a cold raw spring morning, dark, drizzly and gray. Surprisingly the stream was quite full of fish despite the foul weather conditions. Standing by the run I watched as the old woman shuffled down the steep incline toward me. In one hand she clutched a cane, her other arm was intertwined with her granddaughter. The old woman leaned heavily against the younger for support. She was wrapped in a heavy black over coat that seemed to swallow up her hunched over frail figure. Her light blue eyes sat deep in her furrowed face, her complexion was as pale and gray as the dismal morning. As they approached they paused at the bench that sat several feet back from the run. The young woman motioned to the bench, the older woman said nothing. Nodding "no", the old woman now took the lead, shuffling to the river bank. I smiled and said hello as they passed me, the young woman returned the greeting along with a smile. The older woman nodded as if to acknowledge my greeting but said nothing, her face showing no emotion.

Arriving at the river bank, the old woman looked down into the water at the swarms of alewives milling about at the entrance of the run. She then glanced down stream at my children, who were scrambling along the rocks laughing and grabbing at the passing fish. A bit of color came to her face as she looked out on the scene with a far away look in her eyes. I wondered to myself what she was seeing? Perhaps it was herself as a young girl, doing the same as my children were. Or perhaps she was seeing her own children playing on Nemasket's stage. Whatever it was that she saw it seemed to thaw the chill of the morning and lift the burden of old age from her shoulders. When she turned to leave, she looked up at me with a sparkle in her eyes. Then with a hint of a smile she said "yes, it is a good morning young man, a very good morning."

While watching her shuffle back up the incline I couldn't help but wonder how many times similar scenes had been played out here. It's an interesting thought to contemplate, considering Nemasket's long history: 8,000 years ago when the first clay pots were fired in China and the first bit of cloth was woven in Europe and the Middle East, people came here to Nemasket; 4,500 years ago when the first written language was established in Sumeria, people came here to Nemasket; 2,000 years ago during Biblical times, people came here to Nemasket. How many old Wampanoag women have shuffled down to this very spot to relive scenes of their youth? How many fathers, mothers and children have come to this very spot over the past ten thousand years to celebrate the return of the Alewives? How many, I do not know. However I do know that I along with many others find a strange comfort here in the riffles of Nemasket, the place of fish, the river of smiles.

Alewives and the Web of Life

For reasons I can't fathom, crows love to chase and harass bald eagles, nipping at their tails and wings, taunting them like your annoying sibling might have done, hundreds of feet in the air, seemingly just for the heck of it.

One June day in downtown Augusta, an osprey had just done a kamikaze dive bomb into the middle of the Kennebec River, made a huge splash, came out of the water with an alewife and lumbered back into the air to haul the wriggling, struggling fish to her nest of baby osprey in a power line pole in a pasture a mile up Cony Street towards the town dump.

Then a bald eagle took off from a riverside elm tree and did what bald eagles do to ospreys with fish in their talons -- he tried to steal the alewife by attacking the osprey and making the osprey drop the alewife or risk getting pummeled by the much larger (but still juvenile) eagle.

The osprey, though much smaller and burdened with a 14 inch fish and still soaking wet, had no intention of giving up her hard fought prize to a big lazy thief. So the bald eagle and osprey began doing cartwheels over the parking lot of the Augusta post office next to the river. Then a crow, who apparently had nothing much better to do that afternoon, decided this was a great time to harass the bald eagle as it harassed the osprey.

So in short order a three ring circus only seen since the Flying Wallendas began above the Augusta post office parking lot, with the alewife trying mightily to extract herself from the osprey's talons, the osprey trying to hold onto the alewife, the bald eagle dive bombing the osprey to make her drop the alewife, and the crow having free rein to harass the eagle since the eagle was totally focussed on harassing the osprey to drop the alewife.

Finally, after about five minutes of this tussle, which resembled a fight over the last beer in a trailer park, the osprey won, the alewife lost, the eagle lost and the crow won. All that was missing was a song sparrow chasing the crow and a mosquito chasing the sparrow.

Pleas in the 1990s to save the Amazon Rain Forest from clearcutting because it might contain plants which might lead to a cure for cancer always struck me as misdirected, since this premise suggested that if no cancer-curing plants were found, it was okay to level the place, and if the curative chemicals in the plants could be synthesized in a laboratory, it was also okay to level the place. So, either way, this plea was still admitting a license to level the place.

There is ample justification for restoring alewives purely for their own sake, just as we have done with bald eagles. And there is so much evidence of important benefits to other species from alewife restoration that it would be foolish not to save them. But saying you're saving them for reasons other than saving them has always struck me as a loser's bet.

Alewives are prodigious moms and dads because they have to be. Virtually everything eats them at every moment of their life, from their spawning ponds, their migration routes up our rivers, the estuaries where these rivers interfinger with the sea, the cold open ocean gyre of the Gulf of Maine and the treacherous shoals of Nantucket Sound. As Phil Brady, a Massachusetts fishery scientist said to me from Pocasset one day, "An alewife's lot in life is to be eaten."

Where alewives are temporarily obstructed, say at a fishway or natural falls, mother mink and river otters can be seen fishing for them. Great blue heron and snowy egret stand like sentinels at point bars angling their necks in wait for an errant adult or baby swimming too close to shore. Dozens of double breasted cormorants, black and slick as new tires, dive into the deeper pools, come back up and gulp alewives like fat relatives at a family reunion.

Osprey wheel overhead, hover like helicopters and smash dive into the schools. Bald eagles wait for an osprey to catch an alewife and mercilessly chase her hoping to force her to drop her catch so the bald eagle can retrieve it on the wing. A 1700s account from Worromontogus Stream, just down the Kennebec from Augusta, describes black bear and 'swine' feasting on alewives in the spring.

With the return of seals to New England after being wiped out by 19th century seal hunters, we can now see that seals are also avid alewife eaters. They are now sometimes seen on the Kennebec as far upriver as Waterville, Maine, 60 miles from the salt, eating alewives.

Beneath the water, predator fish like striped bass and eels in freshwater, bluefish, stripers and squeteague in the salt, and Atlantic cod (when they used to be common inshore) prey upon the schools. Although observations offshore are quite difficult to make, it's safe to assume nearly all of the fish-eating fish of the New England coast take a whack from time to time if a school of alewives swims by. Watching 70 foot long humpback whales on Stellwagen Bank, there seems little doubt they will devour entire schools of juvenile alewives if and when they can.

When I began working to restore alewives in Maine in the early

1990s, Maine's state agency in charge of alewife restoration, the Maine Dept. of Marine Resources, was often hard-pressed at public meetings to give a clear reason why they were trying to restore alewives. Sometimes the only clear answer they gave was that alewives make good lobster bait.

For a small crowd of rural, inland folk from a tiny Maine farm town 80 miles from the coast, many of whom were already skeptical about 'big guvmint,' the response to someone from the government talking about putting these 'foreign' fish in their local pond was not unpredictable. The next day, hand-written petitions would show up scotch-taped to the counter of the local general store asking town folk to 'stop the alewives,' since it seemed the only beneficiary was some guy hauling lobster pots three hours away on the coast who wanted cheap bait. Clearly, native alewives needed a better P.R. firm.

In 2000, I came across a number of letters from the 1870s by a pioneer U.S. government fisheries scientist, Spencer F. Baird. In these letters, Baird put forth the hypothesis that the sudden disappearance of inshore Atlantic cod populations around the Civil War might be due to the nearly complete damming of Maine's large rivers in the 1830s and 1840s and the ensuing collapse of the rivers' enormous runs of alewives, blueback herring and American shad.

The collapse of the herring and shad runs during this period was well documented and unequivocal, as was the collapse of the inshore cod fisheries during the same period. As a scientist, Baird wondered if there might be a causal connection. Baird knew that the fishing techniques used for cod at this time, handlining with bait (often cut-up alewives) from small dories, catching one cod at a time, had not changed for several centuries, and the fishing intensity had not greatly increased either.

Baird theorized there was no way the crews from Gloucester and the Isle of Shoals and Maine could ever 'fish out' the massive inshore cod schools with drop lines and single hooks. But the disappearance of cod from the near coastal waters during his tenure of observation was undeniable. What had happened? What had changed?

Baird theorized that these billions of herring, born dozens or a hundred miles in the interior, were a principal food supply for the cod, halibut, haddock, pollock and other fish once common along New England's coast. When large, totally impassable dams were built (all illegally) on the St. Croix, Penobscot, Kennebec, Androscoggin and dozens of smaller rivers in the 1820s, 1830s and 1840s, they wiped out the herring and shad and salmon runs and the inshore cod stocks were deprived of their principal and almost limitless source of sustenance, like a squirrel deprived of nut trees or a monarch butterfly deprived of milkweed.

If Baird had paid a visit to the Maine Archives in Augusta, Maine in the 1870s and spent a few hours going over obscure, forgotten documents in a file labelled "Legislative Grave Yard," he might have found a short but urgent entreaty from the citizens of Gouldsborough, Maine to the Maine Legislature in 1824, who wrote:

"The Petition of the subscribers, inhabitants of Gouldsborough in the County of Hancock, humbly represents.

That the Stream emptying into Prospect Harbour in said town, called Prospect Stream, was formerly visited, in the proper season, by great quantities of Alewives, which used to go up said stream to a pond at the head thereof, and there cast their spawn -- that for a number of years past their passage up said stream has been obstructed by a mill-dam erected near the mouth thereof, so that few if any Alewives now pass up said stream -- that in consequence of the obstruction aforesaid they have now mostly forsaken said Harbour and Stream; greatly to the injury of the Cod-fishery on the neighboring coasts; as it is well known that the Cod follow the alewives, in great numbers, even into the Bays and Harbours where they frequent -- that a convenient and sufficient passage for said fish may be made through or around said dam at a small expence, and without material injury to the Mills situated thereon.

They therefore humbly pray your Honours to pass an Act for opening said Stream, and establish such regulations on the subject as wisdoms shall judge proper and expedient. As in duty bound will every pray.

Gouldsborough, Dec. 20th, 1824."

In the 1870s, scientists like Baird did not know if North Atlantic cod were one giant, genetically fluid population or a set of reproductively distinct and separate populations, especially between 'inshore' populations hard on the coast and 'offshore' populations out 100 miles at Georges Banks.

Research today, spearheaded by Ted Ames of Stonington, Maine, appears to conclusively show the latter is true. As with Atlantic salmon, the complete loss of a localized stock (say in the Kennebec River) does not affect the salmon population in a geographically distant river (say the Miramichi or Restigouche in Canada); and in reverse, due to the salmons' precise homing extinct to their natal river, one could wait forever for a few salmon from the Miramichi or Restigouche to venture into the Kennebec and repopulate it.

From Ted Ames' research, it seems cod are more like salmon and alewives than we thought. Cod evolved along the Canadian Maritimes and the Gulf of Maine into many distinct and separate stocks. Once a local stock is depleted or gone it is very hard, or impossible, for the remaining healthy stocks 50 or 100 miles away to replenish it. This is perhaps why the tidal estuary and 'worm-diggin' spot called "Cod Cove" on U.S. Route 1 in Wiscasset and Edgecomb, Maine has not seen many cod for a century even though cod can still be caught 50 miles offshore in the open Gulf of Maine.

Baird's hypothesis was pinioned on a simple premise. Large fish like 50 pound cod and 100 pound halibut eat fish. If you suddenly erase a few billion fish from the coastal waters of the Gulf of Maine in just a few decades, it's bound to have an effect, and not a positive effect. Either

41

the big fish have to find other small fish to eat or they have to move to another place where there is still an abundance of little fish. In the Gulf of Maine that could only mean much farther offshore.

I photocopied Baird's 1870s letters and distributed them at a few meetings to some degreed fisheries scientists specializing in restoring migratory and coastal oceanic fish. At the time we were in the initial stages of an outlandish proposal, formally approved by nobody, to spend about $50 million of U.S. taxpayer dollars to buy and remove three of the lower hydroelectric dams on Maine's Penobscot River to restore its migratory fish runs. The general reception to Baird's hypothesis was perhaps like Spencer Baird received in 1872: scattered indifference with a dollop of scoffing. That year I was at my brother's for Christmas and he had found a good, scholarly book from 1972 called "The Cod" by Albert Jensen. I noticed a brief passage which read:

"Cod particularly go after the schooling fishes -- herring, menhaden, alewives -- and in the northern part of their range, around Newfoundland, for example, cod voraciously chase capelin (Mallotus) ... Each spring, hundreds of cod weighing up to 35 pounds are caught from the banks of the Cape Cod Canal in Massachusetts. The cod pursue the springtime schools of spawning alewives, a small herringlike fish, and are caught by anglers using alewives for bait."

Having spent much of my youth along the Cape Cod Canal, I knew exactly where Jensen was referring, the 'herring run' in Bournedale, where the alewives go up to Great Herring Pond in Plymouth each spring to spawn. Albert Jensen's anecdotal reference dated at least to the 1960s, since his book was published in 1972.

However, in the period between the 1980s and 2000 the inshore cod population around Cape Cod had been hammered by commercial trawlers. The only abundant cod left to be found were far offshore at Georges Bank. By 2000 the idea of catching a 35 pound cod near the mouth of the herring run in Bournedale made as much sense as catching a tarpon there.

I presented Jensen's observations to my colleagues in Old Town, Maine as recent and documented proof showing that large inshore cod, when they were formerly abundant, aggressively chased down alewives swimming out of, and into, their natal rivers along the coast. Still, there were rolling eyes. Then I realized how quickly historical memory can be lost. In 1972 Albert Jensen was describing what he considered a commonplace; by 2000 degreed fisheries scientists would not believe it because it was so foreign to their own recent experience.

Thankfully one fisheries scientist, Clemon "Clem" Fay, was independently pursuing a similar path. Clem was the fisheries biologist for the Penobscot Indian Nation in Old Town. Clem was not a Penobscot Indian, or any Indian. Clem grew up in New Jersey and went to the University of Maine at Orono, at the same time I did, and got a degree in fisheries science.

Clem, as a professional fisheries scientist and a subscriber to the peer-reviewed journals of his profession, had recently seen a profusion of papers from the Pacific Northwest describing research showing intimate and important connections between the disappearance of Pacific salmon and the health of terrestrial wildlife in Washington, Oregon, northern California and British Columbia.

These studies, using quantitative methods and data, showed how dependent many terrestrial animals are on these salmon runs, especially because unlike Atlantic salmon, Pacific salmon all die after spawning and their massive carcasses collect in river pools to rot, and in doing so incorporate their bodies as nutrients into the local food web, down to the microbial level.

One day, Clem mailed me a very provocative research paper on the importance of the Pacific lamprey, a close relative of our New England sea lamprey, to the nutrient composition and health of Pacific Northwest Rivers. We then began trading documents, his mostly peer-reviewed papers from the Pacific Northwest, which I had no access to, and me sending him historic scientific material about Maine and New England from the 1800s, which he had never heard of. The unspoken message we got was that, unlike the Pacific Northwest in the 1980s and 1990s, there had been almost no meaningful scientific research about the connections between the terrestrial, river and ocean habitats of New England since Spencer Baird in the 1870s.

For reasons that can only be described as "being a good scientist," Clem became attached to the fusion of Baird's reasoning, recent peer-reviewed papers from the Northwest, and just plain common sense. After all, we both had seen movies as kids showing grizzly bears and brown bears scooping salmon out of rivers at falls. It didn't take a Ph.D. or a peer-reviewed paper to figure out that if the salmon suddenly disappeared, or if the falls were submerged 100 feet beneath a dam impoundment, the bear would have to find food besides fresh salmon.

From this it wasn't much of a leap that if for 10,000 years cod and halibut and pollock had feasted on billions of alewives, blueback herring and shad pouring in and out of Maine's rivers each fall and spring, and if this giant flow of fish suddenly stopped, it might affect inshore cod and pollock and halibut, perhaps negatively. Never mind bluefin tuna and swordfish.

Clem also introduced to me the concept of migratory fish importing from the ocean what is called 'marine-derived nitrogen' into inland riverine and terrestrial ecosystems via their bodies, progeny and fecal product, again through scientific papers based on quantitative research from the Pacific Northwest. This concept has especial significance for Pacific salmon and Atlantic salmon, because their young thrive in steep and rocky headwater streams that are naturally poor in the nutrients that feed the single-celled organisms (benthic algae) that fuel everything else in a stream, from mayflies, scuds, caddisflies, stoneflies, blackflies and mosquitoes.

Nitrogen comprises nearly 80 percent of the air we breathe, but in this form it is not readily available to plants. Specialized plants like legumes (beans, peas) have developed symbiotic relationships in their roots with nitrogen-fixing bacteria which can actually pull nitrogen out of the air and use it to grow.

But for most plants, nitrogen must come from a secondary source, usually in the waste products or decaying bodies of animals. This is why corn plants, which require large amounts of nitrogen, need to be fertilized with fish (such as alewives) or the manure of herbivorous livestock, like cows, to grow.

To those of us who see our local rivers and ponds carpeted with gooey filamentous algae from non-point source pollution or poorly operated sewage treatment plants, it is hard to see how New England rivers and streams are suffering from a lack of nitrogen, phosphorus or other nutrients. In fact, it seems an overabundance of these chemicals is the problem.

But the key difference lies more in the manner of delivery, how tightly knit ecosystems evolve over time and how quickly, like a house of cards, they can topple. A Devil's Advocate could say to me (and once did): "Doug, you advocate for restoration of millions of alewives to Cobbosseecontee Lake based on such things as bringing marine-derived nitrogen into the lake. But Cobbosseecontee is now suffering from too much nutrients due to poor land use. So how can bringing alewives back help and not hurt the lake?"

One answer is that unlike resident fish, alewives leave the lake each year, adults after spawning, juveniles after growing one summer season. The element phosphorus is the chief cause of unsightly algae blooms that plague many formerly crystal clear New England ponds. A superabundance of phosphorus, due to soil erosion from shoreline dirt roads, lawn fertilizers and septic systems leads to a superabundance of single-celled algae in the water. Add the strong sunlight of summer and you get an algae bloom. Hard experience shows that once a pond 'goes green' due to people putting too much phosphorus into it, the effects are nearly impossible to reverse. This is because when the algae die they sink to the pond bottom and the phosphorus in their cells becomes available for the next year's growth.

Baby alewives make their bodies by eating pin-head sized zooplankton like *Daphnia*, which make their living grazing like tiny cows on the single-celled algae of ponds. When the baby alewives leave the ponds *en masse* in the fall, heading to the brave new world of the wild Atlantic, they take their ingested phosphorus with them, removing it from the pond, and breaking at least in part the phosphorus recycling process. In contrast, resident fish, since they never leave the pond, deposit their phosphorus back to the water column. So we can say alewives can help prevent the slide of a pond into an overabundance of phosphorus by taking the pond's phosphorus with them each year as they swim to sea. [2]

If the goal of 'cleaning up lakes' is to restore them to their natural condition, then that condition must include their keystone native species. Alewives are an ancient keystone species of most lakes and ponds of coastal New England, just as much as loons, pumpkinseed sunfish, brook trout and wood ducks. To leave them out is to admit the job is not yet done.

I think of the rivers of New England as a two way street between the deep woods and the deep ocean; or as capillaries connecting your body's extremities to your heart and lungs. In our bodies we know these "streets" as arteries and veins. Arteries carry blood from the heart, carrying needed oxygen to the extremities. Veins carry the de-oxygenated blood back to the lungs for oxygen replenishment.

But in the circulatory system of our planet, rivers serve as both

2 At Sebasticook Lake in Newport, Maine, a lake long plagued by human-caused algae blooms, the town lowers the lake by up to five feet each fall to physically flush as much algae out of the lake as possible. This artificial, stop-gap measure provides the same physical and chemical service provided naturally, and for free, by baby alewives.

arteries and veins. Oceanic plankton cannot on their own swim up the Kennebec River and feed a baby mink. But they can when incorporated into the body of an alewife. And this alewife, having spent three years feasting on plankton in the wilds of the Gulf of Maine, can be caught by a mother mink along the shore of Stetson Pond in deep, interior central Maine, 100 miles from the ocean, so she can feed her babies.

So what happened to all of these animals, the osprey, mink, heron, eagles, cormorants and loons when this enormous supply of food from the ocean was shut off by dams in the 1830s? It is the cod question backwards.

In the 1800s and 1900s people forgot that like they do every day, fish defy gravity. Migrating fish are the vehicles by which oceanic energy is transferred upstream, against gravity, to feed the land. When you shut off the supply of migrating fish by damming a river, you shut off the supply of ocean energy that feeds the great blue heron 100 miles inland; when you shut off the flood of inland energy in the baby alewives that swim 100 miles to the sea, you starve the cod and halibut 100 miles offshore. Everyone is the poorer for it.

We never had to answer the question of what happened to the osprey, mink, heron, eagles, cormorants and loons when the enormous supply of food going into the ocean from our rivers was shut off in the 1830s because at the very same time, all of these animals were being hunted to extinction, and after World War II, were being poisoned to oblivion by DDT.

We couldn't see the effect of old mill dams on these animals because we were wiping them out far faster than than that effect could be noticed. It was like looking for signs of skin cancer in a body riddled with bullets. By the 1960s we saw a flat, featureless polluted peneplain. Rivers devoid of migrating fish, the wildlife who ate them long gone. Oceans devoid of fish and the fish which eat them. All dependent on each other, and that connection between the oceans and the inland, the

47

spruce woods and the salt marsh, the mountain brook and the sandbar. It was if King Solomon had thought the halfway point of a baby was at the joint of the big toe. Or as Chuck Putnam of Rumford said, describing blasting a squirrel at close range with a shotgun as a kid, "All's that 's left was the tail."

So are the connections between alewives and other wildlife deep and profound? Yes. Are they important? Yes. Are they worthy of citation when trying to defend the alewife against extinction? Yes, in the way a big wrench can pound a nail in a pinch.

Blueback Herring:
Alewives who saved a river

Blueback herring first met me in the Quashnet River above Route 28 in Falmouth on Cape Cod in 1980. I was trout fishing in the Quashnet in June, wading in dungarees, when I watched schools of bluish grey fish swim like torpedos between my feet and linger on the gravelly riffles above me.

I had never seen fish like this before. I cast my flies over them, but they ignored me. They were *Alosa aestivalis*, the blueback herring, swimming up to their birthplace in the Quashnet River to give birth.

Blueback herring look *almost* just like alewives but they are smaller and not as deep bodied. They spawn in the riffles of the streams that alewives swim through to get to the ponds where they spawn. It's kind of a college roommate situation.

Not much is known about the blueback herring in New England because very little study has been devoted to them. Even Bigelow and Schroeder, authors of the magnum opus of New England saltwater fish, *The Fishes of the Gulf of Maine*, confessed to know very little about bluebacks. They write:

"Bluebacks and alewives are difficult to distinguish; experienced fishermen who recognize the existence of the two separate fish cannot always tell them apart, so closely do they resemble one another in general appearance. The most obvious external difference between them is that the back of the blueback is definitely blue green, that of the alewife gray green."

On the Kennebec River in Maine, alewives arrive at the head of tide in mid to late April. Bluebacks appear in late May and keep coming through June. The peak of the two runs are about three weeks apart.

Perhaps the most important distinction between alewives and bluebacks is that blueback herring like to spawn in shallow, rocky riffles and rapids while alewives do not. Blueback eggs are coated with a tough adhesive that makes them stick to river rocks as soon as the mother releases them and the father fertilizes them. The eggs look like tiny clear balls the size of pin-heads glued onto the rocks of the stream. In a few days the eggs hatch into tiny, free-swimming larval fish that cascade downstream in the current to deeper pools. In several months the baby bluebacks grow to about 3-4 inches long and migrate down to the sea for three years.

One sure way to tell an alewife from a blueback herring is to take a sharp fillet knife and cut them open from their belly to expose the inside of their stomach, or peritoneum. The inner lining of an alewife's peritoneum is pearly white; a blueback herring's is a sooty grey. I do not recommend this diagnostic method.

Blueback herring are fish of rivers, not ponds, and seem to love to spawn where the water flows just a few inches over a bed of rounded rocks the size of baseballs, softballs and basketballs. They are uniquely suited to the large, wide rivers of Maine, like the Penobscot and Kennebec, where their lower reaches above the tideline spread out into the vast shallow riffles and rapids called 'Rips' by early loggers and boaters. The word 'rips' was invented because at night, if you sleep beside the river, the water going over the riffles sounds like someone slowly ripping a long piece of cloth. These are the places that blueback herring love to spawn.

I'm not sure why bluebacks have always been neglected and ignored even by very conscientious fish observers since back before the

Civil War. Striped bass certainly do not ignore them, since they appear in New England's big rivers, like the Penobscot and Kennebec, just as the big spawning stripers are getting ready to mate and need a hefty bite of sustenance before and after. It is well known on the Kennebec that the time to catch truly big stripers is in early and mid June, when they are hunting spawning bluebacks at dawn.

Like alewives, bluebacks have an insatiable desire to swim upstream and will do anything to get over an obstacle. Many of the photos Tim and I have taken of alewives in the Kennebec River are actually blueback herring, which only became obvious later when we closely examined the photos. The photo on the cover of this book is most likely a blueback herring, with perhaps an alewife in the background, and a big shad off to the side.

The significance of bluebacks was made apparent to me in 1996. At the time the State of Maine and a number of conservation groups were in the midst of a battle royal to convince the Federal Energy Regulatory Commission (FERC) to deny a new operating license for the Edwards Dam at the head of tide of the Kennebec River in Augusta, Maine. This was the first time a state had ever aggressively sought the removal of a large hydroelectric dam solely because of its effect on sea-run fish.

Edwards, built illegally in 1837 without fish passage had been the subject of the ire of commercial and recreational fishermen throughout the 1800s and 1900s. As documents at the Maine State Archives show, people in the Kennebec River valley have been fighting to remove the dam since the day it was built.

While the dam had numerous owners during its 150 years of existence, all seemed to share a particular obstinacy to doing anything to make the dam less damaging. Its last ownership group, a privately held company called Miller Hydro Group from Lisbon, Maine was no different. Ironically, Miller's refusal in the 1980s to even consider putting adequate mechanical fish passage at the dam was the key to its demise.

After a decade of fruitless negotiation, the State of Maine became so frustrated with Miller's petulance that they decided to instead go to federal court if necessary to rip the dam out of the bed of the Kennebec River for good.[3]

In 1991, led by local fishermen, four conservation groups created the Kennebec Coalition, its sole purpose to remove the Edwards Dam. The groups were the Natural Resources Council of Maine, the Atlantic Salmon Federation, Trout Unlimited and American Rivers. FERC was the federal agency charged by Congress to decide whether to give the dam a new 40 year operating license or order its removal. At the time FERC was (and still is) notorious for being aggressively 'pro-dam' and sloughing off efforts to make them at least less damaging, The Kennebec Coalition knew that to even have a chance at winning they had to compile a mountain of evidence so high that even FERC could not shrug its shoulders and dismiss it.

So in 1995 the Coalition sought the help of a little known fisheries scientist from Wolfville, Nova Scotia, Dr. Michael Dadswell, who taught at Acadia University there. Dadswell was charged with describing, in painstaking, irrefutable detail, the biological benefits of removing the dam versus the harm that would be caused by leaving it in place. Dadswell was also asked to scientifically analyze and critique the growing mountain of documents and conclusion made by FERC staff, the dam owner, and other parties in the proceeding.

After several months of work from his office in Nova Scotia,

---

3   The 'modern' effort to remove the Edwards Dam could be said to have begun in 1974 when spring floods ripped a 150 foot wide hole in the middle of the 900 foot wide dam. As later related to me by Tom Squiers, a fisheries scientist with the Maine Dept. of Marine Resources, he and his DMR colleagues, Lewis Flagg and Bob Dow, took the unprecedented step of seeking a face to face meeting with Maine's Governor, James Longley. Granted the meeting, they pleaded with Longley to not allow the dam to be repaired. Gov. Longley, not known for his conservationist tendencies, denied their request and let the dam be rebuilt without even rudimentary fish passage. The Edwards Dam remained impassable to all fish for another quarter century until it was finally removed in July 1999.

Dadswell delivered to the Kennebec Coalition his 90-page analysis, which later became known to all involved as "The Dadswell Report." Reading it in 1996, after being first knocked over by how insightful and well done it was, I noticed one small, critical comment buried in its pages.

Dadswell noted that in all the documents he was asked to review by FERC, the dam owner, Maine's Dept. of Marine Resources not a single one even mentioned the existence of the blueback herring in the Kennebec River. And, Dadswell wrote, his analysis showed the blueback herring would probably benefit more from the dam's removal than any other fish. To prove his point, Dadswell had to tell the story of the blueback, and oddly enough, sunlight.

Mike's report was (and is) a masterpiece of science working from first principles. Basically, he wrote, this whole hubbub is about how much sunlight can reach the stones on the riverbed. Coastal rivers like the Kennebec have a food web based on benthic algae, the stuff that grows on rocks on the riverbed and makes them slippery. This thin growth of algae on the riverbed rocks is consumed by the tiny larvae of aquatic insects, which in turn feed fish, which in turn feed larger fish, which in turn feed animals like osprey and eagles and us. Because water blocks out much of the visual light spectrum, the deeper the water, the fewer photons of light available to be used by benthic algae. When water in a freshwater river like the Kennebec is too deep, say beyond about 20 feet, benthic algae cannot grow on the riverbed. The food web stops.

Even if you covered the face of the Edwards Dam with fishways of the best and most expensive design, Dadswell implied, the 18 mile long impoundment of the dam would still be there, and that alone would render the river above the dam barren for the all the fish that went up the fishways.

Using surveys of the river from the 1820s, from before it was dammed, and sonar depth maps of the impoundment made in the

1990s, Dadswell concluded that by removing the Edwards Dam the 18 mile reach of the Kennebec above the dam would revert to a broad and quite shallow river. The old surveys indicated a number of large, long riffles, long submerged by the dam with the names like Pettys Rips, Carters Rips, Six Mile Falls and Bacons Rips. With the dam removed, these places would revert to the 'sweet spot' of depth and current ideal for benthic algae growth, aquatic insect productivity and fish habitat. With the dam left in, the water would be too deep to support much life in them at all.

Lastly, Dadswell noted, the wide, shallow, massive riffles and runs which the old surveys showed the Kennebec once had above the dam are the exact type of habitat required by blueback herring to spawn and grow. And given the sheer acreage of these riffles and runs, he predicted that bluebacks would quickly become extremely prolific once the dam and its impoundment were removed. Today, in 2010, a decade after the Edwards Dam was removed, the bluebacks have proven Mike Dadswell was right.

The earliest hand drawn maps of the Kennebec, dating to the mid 1700s, show a place on the river 10 miles above tidewater called "Negwamkeag." dotted with gravel bar islands. With the Edwards Dam removed, these features have come back to life. Here the Kennebec is nearly a half mile wide and very shallow, flowing quickly over a vast bed of softball-sized stones that pile themselves up in a maze of very low gravel bars, flooded when the river is high but prominent in the summer when the river is low. This part of the Kennebec is "textbook" habitat for spawning blueback herring and the sheer amount of it is astounding. For many years now I have gone up to Negwamkeag in early June, and exactly as Mike Dadswell predicted in 1996, the riffles and bars are filled with spawning blueback herring, the males chasing the females around at river's edge and the osprey and heron and egrets trying to eat them.

Blueback herring also taught me how dangerous and damaging it can be when otherwise well-meaning folks try to 'control' the water coming out of a natural lake at a dam. Cobbosseecontee Stream is the outlet of three very large natural lakes west of Augusta, Cobbosseecontee, Anabessacook and Maranacook. The water from the lakes flows into a stream which enters the Kennebec in downtown Gardiner, Maine. The lowermost mile of Cobbosseecontee Stream in Gardiner is not dammed (the rest is) is about 80 feet wide and consists of solid Class II-IV rapids.

In the mid 1990s I started to spend quite a bit of time poking around Cobbosseecontee in downtown Gardiner in part because it is so beautiful and close to my house. One year, around Memorial Day, I noticed hundreds and hundreds of what I thought were alewives charging up the stream. But they behaved oddly. Instead of just going right up to the first impassable dam a mile upstream, these alewives congregated in groups in the many riffles and chased each other around for hours. Having recently read Mike Dadswell's report, I wondered if they were blueback herring. Wading out into the riffles where the 'alewives' were chasing each other around, I picked up a couple rocks which had been right below them. They were covered with tiny, transparent globules that suspiciously looked like eggs. So I put one rock in a bucket of river water, brought it home and put it in my little fish tank. Several days later the 'eggs' hatched into tiny, quarter inch long fish. A few days later they were big enough that I could tell they were some kind of herring and I drove back down to Cobbossee and let them go. They were blueback herring.

A couple days later, energized by my discovery, I went back down to Cobbossee just about Memorial Day. But something was wrong. The stream was about half its size from a few days before. Rapids that had been raging were no a meek trickle. Gentle riffles, off to the side of the channel, were now dry as bones. Most of the places where I had seen bluebacks spawning were now empty of water, and when I picked up the rocks I could see hundreds of tiny eggs clinging to the rocks, but now whitish instead of clear, and obviously dead. After some calls I found that of all things a conservation group was doing the killing. The Cobbossee Watershed District (CWD) was created by the

Maine Legislature in the 1970s to 'manage and protect' the Cobbosseecontee watershed. It has a small professional, scientific staff and is funded by annual contributions from the towns in the watershed. As I learned from calling them, every Memorial Day the CWD shut off most of the water flowing down Cobbossee Stream to keep the big lakes at a carefully specified 'summer recreation water level.' This year, in 1997, Cobbosseecontee was flowing at about 350 cubic feet per second (cfs) until Memorial Day, then within a few minutes, the CWD staff shut the water off at the lake outlet dams, bringing the flow down to about 100 cfs in a few minutes. The blueback herring in Cobbossecontee Stream did not know of this policy, and due to this lack of information, had deposited most of their eggs on sections of the stream that one day were well watered and the next were high and dry.

Since I was probably the first person to notice blueback herring spawning in Cobbossee Stream, this was news to the Cobbossee Watershed District, who had never even heard of *Alosa aestivalis*. But their response to the tragedy was quite odd. First they said they didn't believe me. Then they said that even if they had killed hundreds of thousands of blueback herring eggs, it was not their problem, since technically, their legal jurisdiction does not extend to the reach of the stream where I had been walking. This confused me. They admitted they had caused the drastic reduction in stream flow over Memorial Day which dewatered the lower stream and killed the blueback herring eggs. But then they said that because the killing occurred outside their legislative jurisdiction, they were free to keep doing it.

Henry Thoreau, Alewives
and what he left out.

Hank Thoreau, as my dad would have called him (or maybe "Quahog" or "Spider"), was quite an odd duck. Growing up we knew lots of misfits, spendthrifts, misers, malcontents, non-bathers, raccoon chasers and miscreants in our little town of Easton. People who seemed unable or unwilling to conform to whatever were the acceptable norms as defined by my Aunt Millie and Helen Holster, who kept track of these things. Henry Thoreau, for his time and place, in Concord, was definitely an odd duck.

Since 1996 I have tried to track down any scrap of writing by anyone who was actually there when the first big dams were built on the rivers of Maine and Massachusetts in the early 1800s. These accounts are very hard to come by except for legal documents and petitions to enact laws, most written in dry and stilted legalese. Poring through the Maine State Archives I wondered, did *anyone* in New England ever think to write down what they really thought, felt and saw when this apocalypse occurred?

Then, as usual, my brother Tim answered the question by sending me a few pages from Henry Thoreau's first book, *A Week on the Concord and Merrimack Rivers*. Contrary to popular belief, Thoreau did not go to Walden Pond to write his book *Walden*. He went there to write *A Week on the Concord and Merrimack Rivers*. Ostensibly about a 13-day boat trip he took with this brother John in 1839, when Thoreau was 22, *A Week ...* was written six years later, around 1845-1846.

A Week ... sold only 220 copies, with the publisher sending back to Thoreau the 700 odd unsold copies. In *Walden*, Thoreau jokes about this, noting the shelves of his new 10 x 15 foot house at Walden Pond were well stocked with more than 700 volumes, most of which he had written.

Like everyone else in New England during his lifetime, Henry Thoreau watched as his rivers were dammed and the fish life that once filled them were driven to extinction. But as best as I can tell, very few people besides Henry Thoreau ever put their real thoughts and emotions about this on paper. His were blunt:

"Shad are still taken in the basin of Concord River at Lowell, where they are said to be a month earlier than the Merrimack shad, on account of the warmth of the water. Still patiently, almost pathetically, with instinct not to be discouraged, not to be reasoned with, revisiting their old haunts, as if their stern fates would relent, and still met by the Corporation with its dam. Poor shad! where is thy redress? ... Armed with no sword, no electric shock, but mere Shad, armed only with innocence and a just cause, with tender dumb mouth only forward, and scales easy to be detached. I for one am with thee, and who knows what may avail a crow-bar against that Billerica dam?"[4]

At this time, Henry Thoreau was in his mid 20s and not, as my dad would say, gainfully employed. In today's parlance he would be labelled 'underachiever.' In Walden he goofs on his own lack of ambition, declaring he has appointed himself "Inspector of Snowstorms" for the Town of Concord.

After finishing A Week ... and entering in his journal many of

---

4   What I find interesting about these passages is Thoreau begins as the carefully observant naturalist, notes a distinction in the run timing of shad on the Concord and the mainstem Merrimack, presents a plausible explanation (the Concord run is earlier because its water warms more quickly in the spring), and then quickly slides into a rant of impassioned outrage. And unlike any writer of his time (and few since), he looks at the shads' dilemma through their own eyes. Be the shad !!!

the words he would shape into *Walden*, Henry took three trips to Maine to see what he considered the last piece of true untrampled wilderness left in New England, the headwaters of the Penobscot and Allagash Rivers.

These trips are described in his book, "The Maine Woods." His first trip was in September 1846 in the midst of his 'experiment' in simplified living at Walden Pond. In his story of the trip, titled "Ktaadn," he writes: "Within a dozen miles of Bangor we passed through the villages of Stillwater and Oldtown, built at the falls of the Penobscot; which furnish the principal power by which the Maine woods are converted into lumber. The mills are built directly over and across the river. Here is a close jam, a hard rub, at all seasons; and then the once green tree, long since white, I need not say as the driven snow, but as a driven log, becomes lumber merely. Here your inch, your two and your three inch stuff begin to be, and Mr. Sawyer marks off those spaces which decide the destiny of so many prostrate forests."

The sparse word count Thoreau allots the mill towns of Orono and Old Town suggests he found little of interest there and like any tourist-writer with a destination in mind, was saving his paper and ink for the real story: a long hike to the still wild upper Penobscot culminated with climbing the massive, mile high face of Ktaadn.

This was to be his "A Week On the West Branch Penobscot and Katahdin" and it seems he felt the story did not really begin until he was far above the smell and racket and pell mell of the lumber mill towns of Orono and Old Town. But in an uncharacteristic lack of writers' instinct, Thoreau did not see the real 'story' of his story was centered in Orono and Old Town, since this is where the place he wanted to visit was being choked and starved by giant timber crib dams owned by 'cunning lawyers' working for 'soulless Corporations.' Or perhaps Thoreau just wanted to go on a vacation.

If Thoreau had spent a few more days in Old Town, tarried about, walked to the river, stopped in an inn and asked a few questions, he might well have bumped into Benjamin Shaw, the Penobscot County fish warden who at that moment was trying his hardest to bring back the 'finney millions" of alewives and shad to the Penobscot by enforcing the state's fishway laws. Surely in Shaw Thoreau would have seen a kindred spirit.

"Ktaadn" describes in valuable detail many sites and places along the river from Old Town to the upper West Branch. He notes seeing flocks of passenger pigeons and several 'fish hawks' (osprey) along the river and an eagles' nest in a riverside pine so big it could be seen from a mile away. He correctly notes the habit of lazy bald eagles trying to steal on the wing the fish that osprey have just dove for and caught. Staying for a night at Shad Pond in Millinocket near the mouth of the West Branch he notes a local eating the "the last salmon caught that season" but never ruminates on how a place that far upriver got the name Shad Pond and why it was bereft of shad.[5]

For such an astute observer of nature and wildlife, it seems totally out of character for Thoreau to have not commented even once on the native fish fauna of the Penobscot, especially since its key species, salmon, shad and alewife and eels, were the same as on his own Concord and Merrimack Rivers. During their excursions together, Thoreau repeatedly quizzes his Penobscot guide, Joe Polis, about the Penobscot names for various birds, but not a whit about the Penobscot Indian names for the fish of their own river or what happened to them.

To be fair, several factors worked against Thoreau on his trips to Maine. First, he went there in September, months after the shad, alewife and salmon runs of May, June and July had concluded. Second, by 1846 few sea-run fish except a small number of salmon could get above the

---

5   "As we stood upon the pile of chips by the door, fish-hawks were sailing overhead; and here, over Shad Pond, might daily be witnessed the tyranny of the bald-eagle over that bird. Tom pointed away over the lake to a bald-eagle's nest, which was plainly visible more than a mile off, on a pine, high above the surrounding forest, and was frequented from year to year by the same pair, and held sacred by him." From "Ktaadn."

gauntlet of dams in Veazie, Orono and Old Town. Thoreau's most detailed description of the Penobscot (in the essay "Ktaadn") is devoted to the upper West Branch above the recently built and impassable dam at North Twin Lake near Millinocket. Although he was camping beside rapids once teeming with salmon just a few years earlier, Thoreau's text suggests he was only vaguely aware of it. Lastly, since it appears these Maine essays were first intended for magazine publication, it might have been that Thoreau felt a full-out rant against dams might doom its publication chances.

That said, "Ktaadn" is not without moments of his bone dry, yet scalding sarcasm. After observing massive log booms and saw mills on the river for miles, he says: "Think how stood the white-pine tree on the shore of Chesuncook, its branches soughing with the four winds, and every individual needle trembling in the sunlight, — think how it stands with it now, — sold, perchance, to the New England Friction-Match Company!"

Yet we know, through numerous historic documents included here, that when Thoreau was on the Penobscot, the river valley was awash in bitter controversy about the effect of the recently built dams at Veazie and Great Works and Old Town and the dam owners refusal to keep the river passable to fish. But to read "The Maine Woods" you would never know it. Why was Thoreau so mute and incurious? How was the young man who talked of taking a crowbar to the Billerica Dam because it stopped the shad so uninterested in the same thing happening on the much larger Penobscot, even as he waded in it?[6]

---

6   Two years before Thoreau's trip, a legislative petition from the Town of Bucksport stated in very Thoreau-like language: "There are two dams only that obstruct the free passage of salmon, shad and alewives up the Penobscot river, 'The Corporation,' owned by John Otis and others, and the 'Great Works,' owned by Josiah S. Little and others. These gentlemen lawyers have had the cunning to evade for years that portion of the law as it stands in the Revised Statutes which requires them to open a fish-way by their dams, and it is found insufficient to effect that purpose. The law of the last year will compel these soulless Corporations to open a passage way for said fish, which can be done without danger to those structures, and comparatively speaking, at a small expense."

I wish we knew. Thoreau's odd silence in "The Maine Woods" deprives us today of a unique eyewitness account of what it was *really* like on the Penobscot at this critical time. By error of omission, Thoreau inadvertently set back our understanding of the *real* Penobscot River by 150 years. Today, countless people read and cherish "The Maine Woods," believing it is an accurate description of the pristine and primeval Penobscot River. It is not.

When Thoreau visited in 1846, the Penobscot was as denuded of life as his own Concord and Merrimack. The Penobscot he paddled from Indian Island to Millinocket was missing a critical ingredient: its fish. Nearly all of the native fish of the main-stem Penobscot, in number and weight, are fish from the sea: alewives, shad, salmon, blueback herring, lamprey and eels. Take these away, as had occurred by 1846, and the river is still very pretty, but also quite lacking in fish. It is not surprising the only fish Thoreau describes seeing or catching on his trip are fall fish ('cousin trout') and brook trout ('speckled trout') and only in the river's highest headwaters in the shadow of Ktaadn.

Had Thoreau canoed the river in September just 20 years earlier, he would have seen millions of baby alewives, herring and shad beginning to course out of the tributaries to the sea, doing their 'happy dance' of crescent leaps in the shallows except when chased and attacked by 20-40 pound native striped bass.[7] Shad Pond, a wide spot on

7 Early 19th century newspaper articles collected in a centennial history of Bangor, Maine from 1872 describe a massive forest fire that devastated the middle Penobscot River and Piscataquis River valleys in August of 1825, before any significant dams existed in the Penobscot River's mainstem. These accounts describe very large (20-40 pound) striped bass in the Piscataquis River: " For a fortnight fires were raging in the forests north of Bangor. At one time nearly the whole country from Passadumkeag to Mattanawcook, on both sides of the Penobscot and Piscataquis, was a sea of flame. The roaring of the fire was like thunder, and was heard at a distance from twelve to fifteen miles. The islands in the river were burnt over. The country between Passadumkeag and Lincoln was devastated. The towns upon the Piscataquis suffered from loss of buildings, cattle, fences, crops. The house, barn filled with hay, and store and toolhouse of Joseph McIntosh, of Maxfield, were burned and the family driven to the river for safety. Other houses and barns, and saw-mills and grist-mills, were

the West Branch where he camped, would have been occupied by hundreds of thousands of baby shad with four foot long American eels greedily devouring them as they prepared to go to the Sargasso Sea. Thoreau saw and wrote about none of these sights because he couldn't. The damage had already been done by the dams and mills he casually noted on his short ride through Orono and Old Town.

What I glean from Henry's protests about the devastation of the Concord and Merrimack is it is inconceivable only he had this reaction. He was just one of thousands and thousands who witnessed the dams built and the fish driven out of their own backyards. Even if most people of the time were apathetic or complacent, still a few would have been, like Thoreau, enraged enough to put their words on paper. Who were they? Where are their stories? Where are they preserved? Who heard the fishes when they cried? The documents here, most never seen since the year they were written, at least tell us Thoreau was not alone.

---

destroyed. A lad returning from school through the woods was so badly burned that his life was despaired of; hawks and other birds were killed by the fire; and the fish in the Piscataquis River were killed by the heat. Twenty bass, weighing from twenty to forty pounds, many young salmon, shad, trout, and other small fish, were found dead in the shoal water and on the shores." This is the only existing written account of striped bass on the Penobscot above Old Town. Despite this eyewitness account, the 'conventional wisdom' of modern fisheries biologists in Maine remains that striped bass 'could not' go up the river past the rapids at Old Town. Nobody has ever explained why.

The Curious Case
of the Cobbosseecontee

The history of the alewife in New England in the last quarter millennia can perhaps be summed up best by the case of Cobbosseecontee Stream, which enters the tidewaters of the Kennebec in south central Maine.

The name means 'sturgeon catching place,' not the stream itself, but the waters of the Kennebec at its mouth at what is now the small city of Gardiner, Maine. [Cabasa = sturgeon; Contee = catching place; place of abundance]

Cobbosseecontee (shortened by locals to Cobbossee) is about 200 square miles in watershed size, much of it occupied by natural lakes, all draining into a medium sized meadowy stream until it reaches the outskirts of Gardiner, where it drops 125 vertical feet in just a mile and a half to meet the Kennebec 35 miles from its mouth at the Atlantic Ocean.

The large lakes of the stream's headwaters were all originally inhabited by alewives, making Cobbosseecontee, acre for acre, one of the largest habitats of the species in New England. By contemporary estimates of alewife production, the lakes and ponds of the stream annually supported 4-5 million alewives swimming upstream each spring and perhaps as many as 100 million juveniles pouring down its course to the ocean each late summer and fall.

To put this into perspective, there were more alewives in Cobbosseecontee in 1750 than in all of New England's rivers today.

No descriptions exist of how this sight must have struck the first English settlers when they built, as their first course of business in 1761, a large log dam at its mouth which quickly destroyed the run.

Records show they were workmen hired by a wealth Boston pharmacist and real estate speculator named Silvester Gardiner, who had acquired development rights for hundreds of square miles of land in central Maine from the Kennebec Proprietors, a much larger group of Boston land speculators which included James Bowdoin, the founder of Maine's Bowdoin College.

The impetus to quickly 'settle' the Kennebec lands in the mid 1700s had international purposes tied to England's long battle with France for control of North America, wherein Maine's Indians, especially those on the Kennebec, had long sided with the French, and by this allegiance were viewed as enemies of the English Crown who needed to be subdued and ultimately removed by force. To secure these lands required settlers.

To begin his development plan, Silvester Gardiner hired a small crew of workmen and a dozen or so willing Massachusetts families to stake out a permanent outpost at the mouth of Cobboseecontee, to be named, naturally, Gardinerstown. They made shipfall at the stream in the summer of 1761. Needing houses to live in, they began cutting the giant white pine in the area for lumber but needed a way to turn the logs into lumber; so a dam was built at the stream's first falls and equipped with a rudimentary up and down saw. From this contrivance, saw logs could be made and turned into houses for the coming winter.

While records are brief and scant, it appears that in the spring of

1762, the 4-5 million alewives which had surged up Cobbosseecontee each year for the past 8,000 years were suddenly stopped by an 8 foot wall of stack of stout pine logs pinioned across the stream channel. The sight of this struggling, thrashing mass of millions of fish trying in vain to get to their birthplace was apparently not worth mention by those first settlers and hired workmen.

From this date, and for the next 250 years, the descendants of those families, and those who came long after, were equally incurious, with the latter having no clue or inkling to what had used to be. Today, not a single historical account of Gardiner and its environs contains a mention of the fish of the alewives of Cobbosseecontee. In just a couple of seasons they had been swept down Orwell's memory hole.

But this sweeping clean did not reach every recessed corner and alcove of the house of history. The impending outbreak of the Revolutionary War caused some local settlers to reconsider whether destroying the giant alewife runs of Cobbossee was a good idea after all.

By time of Lexington and Concord, Silvester Gardiner, a committed Tory living in Boston, had been driven to England as part of the city's Tory evacuation. The massive British naval fleet patrolled the waters of Maine, blockading the coast and intercepting any ships or boats carrying people or provisions. The tiny settlement of Gardiner and the even smaller outposts which had sprung up along the headwater lakes of Cobbosseecontee were shut off from contact with the rest of the colonies to the north and south. The disastrous naval battle in the Penobscot at Castine, where much of the nascent Continental Navy was burned and sunk gave the English total supremacy of the New England coast and put a seige on tiny towns along the shores of Maine. One such settlement was at Pondville, which is now the small bedroom town of Winthrop, Maine.

Records show that one of the first official votes at the first town meeting ever held in Pondville, in 1771, was to appoint a committee of men of the town to meet with "Mr. Gardiner" to compel him to allow passage for alewives up the 15 miles of stream from the Kennebec to the ponds of Cobbosseecontee. Gardiner, through his son William, who had been left in charge of Gardiner's lands while his father took refuge in London during the war, refused the requests of the people of Pondville to let the alewives come upstream. Winthrop historian David Thurston wrote in 1855:

"The first action of the people at Winthrop in relation to the dam, on record, is at a meeting of the town, Nov. 17, 1771, when they chose James Craig, Jonathan Whiting, and Ichabod How, a committee to solicit Dr. Gardiner & Son to open a place through, or around their mill dam, to let the fish up for the benefit of the town."

And the next year: "In the warrant for the meeting, March, 1772, the 5th article was, 'To choose a committee to solicit Mr. William Gardiner to open a place through or round his mill dam, to let the fish up for the benefit of the town.'

And the next year: "August 30, 1773, 'The committee made a verbal report to this purport, that they had waited on the Dr., and desired him to open a suitable way through or round his mill dam, for the fish to go up for the benefit of the town, but that the Dr. wholly declined to comply with their request.'"

And two years later: "July 10, 1775, chose Joseph Baker, Ransford Smith and John Blunt, a committee to obtain a fish way through Mr. Gardiner's mill dam at Cobbossee in some lawful way."

On June 29, 1776, a few days before the Declaration of Independence was signed in Philadelphia, the people of Pondville took their grievance to the provisional Continental government in Boston via a petition which read:

"To the Honorable the Council for the Colony of Massachusetts Bay and the Honorable House of Representatives of the Same in General Court Assembled The Petition of Joseph Baker, Ransford

Smith and Daniel Dudley a Committee of the town of Winthrop in the County of Lincoln in Said Colony in behalf of the Town Humbly Sheweth ...

"That Said Town is Situated in the River Called Cobiseconte formerly noted for one of the best streams in these parts for Fishing but some years ago Doct. Silvester Gardiner late of Boston Erected a mill dam at the mouth of Said River where it empties into the River Kennebeck which entirely stopped the Course of the fish up Said River called Cobiseconte. The Inhabitants of Said Town Sensible of the Great advantage of the fish taken so near as they might if they were not stopped by Said mill dam applied to Said Doct. Silvester Gardiner to make a fish way through or round his mill dam which he seemed willing at first to do but after delaying from one time to another refused to do anything about it and the Town having no other way to obtain a course for the fish up Said river but pursuing the measures printed out by the Law of the land which they have been prevented from taking advantage of by the breaking out of the present Troubles and Considering the advantage the fish would be in case they could have a Course up not only to the Inhabitants of Winthrop but to others in the Neighborhood Your Petitioners pray your Honours to take their Case under Your Consideration and Grant Relief by ordering the occupiers of saw mill dam to make a course for the fish by said dam or otherwise as your Honours in your Wisdom shall See fit and your Petitioners shall ever pray."

The daily Journal of the Massachusetts Legislature and General Court, a copy of which resides in the Maine Legislative Law Library, gives a brief and cryptic response to this petition:

"A petition of Joseph Baker, setting forth, That the town of Ransford, is situated on a river called Cobbiseconte, noted for one of the best streams in those parts, for fishing; that Sylvester Gardiner has erected a mill on said river, and thereby obstructed the passage of fish up the same, therefore praying that a course may be made in said dam, for the fish to pass. Read, and Voted, That the petitioner have leave to withdraw his petition."

Winthrop town meeting records, as related by historian David Thurston, show that the people of Pondville refused to give up:

"May 17, 1779, they appointed 'Capt. John Blunt, Lieut. Jonathan Whiting and Mr. James Craig, a committee to the Court of General Sessions of the Peace to obtain a fish way round or over Mr. Gardiner's mill dam, at the next session to be held at Pownalborough in June next, and to pursue the affair, at the expense of the town, as they in their judgment shall think best, till they obtain said end, or shall be satisfied it is not attainable.'"

"May 3, 1784, Capt. John Blunt, Robert Page, and Samuel Foster, were appointed a committee to procure a fish way through Mr. Gardiner's mill dam if possible."

"April, 1789, Benjamin Monk, Squier Bishop and David Foster, were a fish committee."

"March 1, 1790, Samuel Wood, Joseph Metcalf and Capt. Nathaniel Fairbanks, were a fish committee."

A close study of the events of this era show a curious set of facts. Masschusetts law dating back to 1736 required mill dam owners throughout the Colony to provide passage at their dams for alewives. This law was strengthened in 1742, apparently because mill dam owners refused to obey the law. This amendment stated in its preamble:

"Whereas, notwithstanding the several acts made for the preservation of the fish, and to give them free passage up and down the rivers, in their seasons, yet by reason of many dams erected, and often erected across such rivers and streams where the several sorts of fish pass up into natural ponds to cast their spawns, said fish are diverted in their passage, to the great decay and ruin of such fishery --"

All of the several sawmill and gristmill dams built at Cobbosseecontee from 1761 forward were built and maintained in clear violation of colonial law since they contained no space for alewives to pass through them. By the same token, the people of Pondville, beginning in 1770 were in the legal right, while "Dr. Gardiner" and his son were flagrantly violating existing law. The colonial Legislature's abrupt dismissal of the town of Pondville's 1776 petition was, in effect, a refusal by the colonial Legislature to enforce a law which had been on the books for 40 years.

From 1770 to 1800 the people of Pondville were not asking that a law be passed to provide them relief but were asking that a longstanding law be obeyed and enforced.

This highly detailed record at Cobbossecontee from the late 1700s illustrates a phenomenon which traces the entire history of New England to the present day: governments have passed endless laws to protect fisheries, incuriously refused to enforce them, and then let those who flagrantly violated the laws get away with it until there were no fish left for the laws to protect.

By the 1790s, Sylvester Gardiner had died and left his estate and holdings and dams to his grandson, Robert Hallowell Gardiner. With the Revolutionary War over and peace and normalcy returning to Maine, Gardiner wished to fully proceed on an ambitious plan to fully develop the 'water powers' of the rushing, raging rapids of the lower two miles or so of Cobbosseecontee Stream.

Unfortunately for R.H. Gardiner, in 1789 Massachusetts passed another, even stronger law to compel recalcitrant mill dam owners to provide passage where their dams blocked runs of alewives.

In 1791 Gardiner decided to finally silence the annual protestations and town meeting actions and 'fish committees' of the people of Pondville. That year he, he petitioned the Massachusetts Legislature to forever exempt Cobbosseecontee Stream from the same fish passage laws his family had been violating for the past 30 years.

But Gardiner didn't argue for this exemption solely on the basis that the value of sawmills outweighed the value of alewives, although he made this claim. He didn't argue solely that making a space for the fish at his dams would be expensive and hard to maintain, although he made this claim. Instead he added a novel twist. Robert H. Gardiner flatly asserted there had never been any alewives or sea-run fish in Cobbosseecontee Stream in the first place. His lawyers wrote:

" ... as the oldest inhabitants in that Country cannot recollect any instance of the Alewives proceeding above the aforesaid Dams, and as a variety of natural obstructions render it highly improbable, that the larger fish would ever proceed above said dams in any considerable number -- "

The people of Pondville responded quickly, producing sworn affidavits from local people who had watched alewives swim up Cobbosseecontee in years past when the Gardiner's dams had been seasonally battered by ice and floods and washed out.

John Stain testified: "I John Stain of Lawful age testify and Say that about thirty years ago before there was any mill Dam built across Cobesecontee Stream I caught Shad fish in said Stream up at the falls about a mile from the mouth of said Stream where a saw mill now Stands and have for years together when I was there to Catch fish Seen Shad and Elwives to over the falls going up said Stream.

Abraham Wyman testified: "Abraham Wyman of Wyman's Plantation in the County of Lincoln, Gentleman of Lawful age, testifieth and saith that some years before there was any mills built on Cobesecontee stream so called which Emptyes in to Kennebeck River at Pittstown. I was hunting on said Stream and I saw a plenty of alewives Runing up said Stream they were then a mile above what was called the upper falls and further the Deponent saith not."

71

Joseph Greeley testified: "The Deposition of Joseph Greeley of Sandey river in the County of Lincoln yeoman of Lawfull age testifieth and saith that about four or five and twenty years ago and to the best of my Remembrance it was the year that Cobboseecontee mill Dam was Caried away I was a hunting on Cobbosseecontee Stream so called that Emptied into Kennebec River at Pittstown and up said Stream at the falls in Winthrop where John Chandler Mills now Stand I Saw a Plenty of Alewives Runing up Said falls. I also Saw Major Heald the same day he informed me that he had also Seen them as well as myself and further the deponant Saith not."

Apparently, this testimony convinced the Massachusetts Legislature to deny Gardiner's efforts to exempt his dams from the state's fish passage laws, but did not convince the Legislature to take action to force Robert H. Gardiner to actually comply with the law. The dams remained closed; the alewives and people of Pondville denied.

Winthrop town meeting minutes show that each year from 1791 to 1796 the town directed the fish committee continue trying to legally force Gardiner to obey the standing Massachusetts laws.

Then, in 1806, the journal of the Massachusetts Legislature records this brief entry:

"Be it enacted by the Senate and House of Representatives, in General Court assembled, and by authority of the same -- That all laws heretofore made, which regulate the fishery of Salmon, Shad and Alewives, in Cobbossee Contee River, so called, within the town of Gardiner, in the County of Kennebec, or that respect any mill-dam across said river, be so far repealed, that from and after passing this Act, they shall cease to operate or have any effect, so far as respects said river, or any part thereof. Approved February 17, 1806."

Robert H. Gardiner, somehow, had finally won. He and his fellow investors quickly proceeded to build nine high stone dams on Cobbosseecontee Stream in Gardiner, named Dam No. 1, Dam No. 2

etc. and lease out their mechanical waterpower potential to dozens of individual sawmill, shingle and textile mill operators along the stream, collecting annual rent from each. In the 1880s, Gardiner's heirs purchased the flowage rights of property owners at the upstream lakes and ponds, allowing them to erect dams raising the natural level of these lakes by ten or more feet, ensuring a steady summer head of water could be delivered to the complex of mills in Gardiner.

By World War I, the industries using the Cobbossee dams evolved from sawmills to paper and textile mills, which dumped thousands of gallons of toxic dyes and sulphite chemicals into the stream each day. Not only could the few alewives in the Kennebec not get past the dams, but the lower pools where a few still vainly tried became a poisonous, anoxic soup. This condition continued until the early 1980s until most of the factories had closed and burned down and the few left were forced by the U.S. Clean Water Act to send their industrial waste to the Gardiner Municipal Wastewater Treatment Plant.

In the 1990s, I talked with a number of people who grew up in Gardiner along Cobbosseecontee in the 1960s and 1970s. One was Patrick Colwell, at the time a freshman state legislator. Pat told me that as a kid in the early 1970s he remembered the stream in Gardiner as being severely polluted. "The water was cloudy and murky and the rocks in the stream were all coated with a weird, whitish grey kind of slime," he said. While nobody fished or went in the stream, he said, kids like himself still explored it. On rare occasion they would still see a few alewives and sometimes, an Atlantic salmon. Others in town, around Pat's age, related to me similar memories of Cobbosee in the 1970s.

By the 1960s, disuse, neglect and floods caused all but three of the nine stone dams built by Robert Hallowell Gardiner's men to collapse into the stream. Since the mills they powered had been closed for decades, there was no interest in repairing them. Their cut granite blocks and stout, handhewn timbers, scattered, shattered and dissassembled by dozens and dozens of spring floods became

73

anonymous parts of the streambed. Maples, white ash and oaks, some now 40 feet high, took root in crevices of the forgotten dam abutments, their leaves forming a blanket over the scrap metal and broken brick of factory buildings that had fallen down a half century prior.

When I moved to Augusta, Maine in 1991 and drove weekly through downtown Gardiner I couldn't help but notice the long, sinuous rapids of Cobbosseecontee as it curved next to Water Street and made its way downtown. It looked deliciously inviting, like the Waits River, a mid-sized trout streams I fished as a teenager in central Vermont. Wading along and through the stream one summer day I noticed that much of its banks cut through what looked like a giant dump. Its bed seemed as much composed of metal and wooden junk and broken glass ande bricks as glacially strewn river stones.

In 1996 I had taken a job as a news reporter for the Capital Weekly, a small, fledgling paper in Augusta, Maine, five miles up the Kennebec. Poking around for stories I spent an afternoon yakking with Lewis Flagg, a fisheries biologist with the Maine Dept. of Marine Resources about fishing and whatnot at his office in Hallowell.

He said, "Doug, if you want to get some nice pictures of leaping Atlantic salmon, go to the first dam on Cobbosseecontee Stream in Gardiner. They come right out of the water."

As I learned, Flagg and his associates had spent a couple seasons at the dam testing a new type of inclined wooden ramp, of a design invented in New Zealand, that appeared an effective way to allow baby American eels to climb over small dams. While installing and testing the ramp at the first dam on Cobbossee, they had noticed 10-12 pound salmon trying in vain to leap the dam, some coming 3-4 feet out of the water.

On Veteran's Day in 1996 I took Lew at his word and clambered across and through and up Cobbossee Stream to the secluded, steep valley where the first dam at Cobbossee was hidden far from the road.

74

Sure enough, within a few minutes I watched as 3 foot long Atlantic salmon, darkish grey on that cloudy overcast day, shot like missiles from the whitewater at the 12 foot dam and were thrown back by the water into the plunge pool. Despite their efforts, their jumps still fell a clear 8 feet from the top of the dam. I shot several rolls of film of them leaping and then made my way back downstream toward my car.

At a deep pool just above where my car was, I saw a man fishing and then saw him filling a hole with his feet in the wet gravel next to the stream. As I got closer and said hi, he said, "I thought for a minute you were a game warden. Let me show you this." Then he bent over and scraped the gravel from the hole he had just filled in the riverbank. As he scraped away the pea stones, the still-wet body of a large Atlantic salmon was revealed. It was a female and was still moving her mouth and trying to breathe.

Even though the guy weighed had 5 inches and 70 pounds on me I dove down to the ground, grabbed the salmon by the tail, hoisted her out of the hole and said in the steadiest voice I could, "Sir, this is an Atlantic salmon. It's illegal to keep them." Before he could speak, I had already thrashed into the center of the channel up to my waist and held the female a few inches under water so the current could pass over her gills. Surprisingly, the man didn't wade out and clock me. Instead, he weakly said from shore, "I thought it was a big trout." By this time, I knew the man knew full well he had caught an Atlantic salmon, not a big trout, and hearing someone walking in the woods who might be a game warden, he hurriedly buried the female alive in the gravel to conceal the crime. This alone demonstrated intent. If he had truly thought he had caught an 8 pound brown trout, which are legal to keep, he would never have buried it upon my approach and then proudly unburied it when he saw I was just another angler, not a game warden.

This calculus shot through my head even as I tried to stand straight in the rapids of the stream, in 45 degree water, holding the salmon with two hands and hoping the current flowing over her gills might revive her. For the next several minutes, all three of us stood as mute as statues, frozen in a weird moment of life, guilt and death. As my hands went numb the female began to move her gills more vigorously, like a heart slowly beating. Her thick tail, encircled by my left hand, started to flex slowly but ever more strongly. Soon she flexed her whole body powerfully and did a U-Turn downstream.

As I waded out of the water, the poacher was still standing there in shock at what had just happened. His last words to me were on the order of, "Really, I thought it was a trout."

Later that fall and winter, after checking with some local anglers, I discovered that a small cadre of poachers routinely fished Cobbosseecontee in the fall and caught "big trout" weighing 8-15 pounds, much larger than any caught in the area. I also found a news clip in the Kennebec Journal archives from several years earlier with an angler displaying a "giant trout" from Cobbosseecontee, weighing about 13 pounds, caught in November. These anglers were poaching Atlantic salmon off their spawning beds in the stream, and I believe most knew exactly what they were doing, which was the only reason the angler I met buried the salmon in the gravel because he thought I was a game warden.

From that day I started exploring Cobbossee much more. The next spring, driving across the bridge in downtown Gardiner to do a newspaper interview, I noticed the stream had gone from Class III whitewater the day before to a bare trickle. Headed back to the office in Augusta, I took a couple quick photos of the dry streambed and headed back to Augusta to file my story. Then I made a couple calls to state agencies, telling them that Cobbossee Stream looked almost completely dried up: in late April.

Lew Flagg, at Maine DMR, took my call and said "they must have shut off the flow at the dams upstream to do some repairs. This happens quite a bit." Dana Murch, hydropower supervisor at the Maine Dept. of Environmental Protection said much the same thing. I asked both what actions they intended to take to stop the de-watering. Neither were very committal. So I wrote a short story for the paper noting the event, which ran on the front page. Soon after, the stream returned to a fairly normal level of water for late April.

Now as a working reporter, this event put me in a pickle. According to the canons of journalism, reporters are not supposed to 'get involved' in a story; they are supposed to be objective, detached observers who report the actions of others. The morning I noticed that Cobbossee Stream had gone in 24 hours from full to nearly dry, thousands of people in Gardiner had driven over the bridge, glanced out their car windows, and saw exactly what I had seen. Nobody had called to complain or even inquire why a raging whitewater stream in downtown Gardiner had suddenly gone dry overnight in April. I was further informed that this was a frequent occurrence that had been going on as long as people could remember.

The next week, a formal meeting was convened in Gardiner with Maine DMR, Maine DEP and the representatives of the various dam owners, including a man in his 60s named John Pulis, the chair of the board of directors of the Gardiner Water District, which owned the uppermost of the three dams in Gardiner. I attended the meeting as both the witness to the event and as a reporter for the Capital Weekly newspaper. After much back and forthing, moderated by Dana Murch, it was established that the Water District had dried up the stream in order to make some repairs at their dam and that, as Lewis Flagg had told me, this was a commonplace occurrence. While removing my name from the story as the complainant, I wrote a very stenographic account of the meeting, including who did it and what was to be done to stop it in the future. Most disturbing was that Mr. Murch told me that although Maine DEP admitted the stream had been almost completely dewatered, because none of the fish in the stream were considered "game fish," but were only native fish like white suckers, the Maine DEP did not consider the de-watering to have damaged anything valuable.

Not long thereafter, Mr. Pulis wrote a letter to the newspaper stating that I had violated journalistic ethics by being both a complainant and the person who wrote the story. I later learned he had also written a private letter to the newspaper's publisher making even more pointed complaints. Confronted by my own editor, I told him I didn't know what to do. If, as a reporter, I saw a house burning down was I not supposed to call the fire department or check to see if there were people in the house? Was I supposed to wait and hope to see if someone else from Gardiner would report that Cobbossee Stream had mysteriously dried up overnight? And if nobody noticed, was I supposed to just pretend it wasn't happening?

Soon thereafter, I wrote a letter, from my home, to all the parties stating that as a fisherman who used Cobbossee Stream, I expected the State and the dam owners to ensure this incident did not happen again. That letter was then forwarded by Mr. Pulis of the Gardiner Water District to my publisher and within several weeks I was fired from my job as a newspaper reporter for the Capital Weekly. When I asked why I was being fired, I was told, "You know why."

From that day, although I had to scramble to find other employment to pay the rent, I was at least free to advocate full-throated for the welfare of Cobbossee Stream. In doing so, I created a tiny non-profit organization, Friends of the Kennebec Salmon, with the help of a long-time river advocate, Bill Townsend of Skowhegan, Maine.

That same month, in late May, my brother Tim came up from Massachusetts and we went down to Cobbosseecontee to the first dam, then called the Yorktowne Dam. It's a sturdy 12 foot edifice of locally quarried granite, fitted into one ton blocks and built before the Civil War. Once used to turn mechanical waterwheels, it hadn't done anything but block fish since World War I. On that day, standing on its crest, we looked in the stream below and saw 20,000 alewives fresh from the Atlantic Ocean, hopelessly beating their heads and bodies against its concrete spillway trying in vain to get over it. Feeling both sorry for the alewives, and angry at the 'system' which allowed it with impunity, we bought a couple landing nets and grabbed some drywall buckets and spent the next two days laboriously netting alewives and carrying them a dozen at a time over the dam. By Sunday, we had moved perhaps five hundred, and even though we gave them access to just a third of a mile of stream before the next dam, we felt a bit better. At least our own impotence had been somewhat alleviated.

The Hunter becomes the Hunted

In the late fall of 1997 I aggressively scouted Cobbossecontee for signs of spawning Atlantic salmon and noticed several. The large nests they dug in the stream gravel in its rapids were obvious to a fairly skilled eye; and when the sunlight was bright and high overhead I could sometimes see the salmon, a few feet underwater, the females digging the nests and the males defending the females against other competing males. Water and light conditions made photography difficult. But I was able to construct a hand-drawn map of exactly where the salmon eggs were buried.

The next April, in 1998, I visited the stream every few days and began to see a few tiny salmon, the size of paperclips, hovering near the salmon nests. The next day, about April 18, the stream flow dropped two feet overnight and the salmon nests were high and dry on raised gravel bars.

After some phone calls I learned the Cobbossee Watershed District had drastically curtailed the flow from all of the headwater lakes because it had been a very dry April and the District wanted to 'conserve' water in the lakes to accommodate boaters and shorefront property owners in case the drought persisted through the summer. I called all the various governmental parties and informed them that least five Atlantic salmon nests, full of hatching eggs, were now de-watered and without the gates being open the hatching salmon would die. A Maine fisheries biologist, Scott Davis, came with me to one of the gravel bars. He dug in with a shovel into the gravel, which was still wet a few inches down, but what we found were not salmon eggs. They were too small. They were white sucker eggs, laid down just a week before.

The next day I returned to the same spot and dug with my hands a couple feet from where he had dug. About six inches down, I gently turned the gravel in the water and a cloud of pinkish salmon eggs the size of peas floated up in the water. Some were cloudy and beige, meaning they had died. But others were still pink and had the tiny black eyes of living Atlantic salmon almost ready to hatch. Others were 'yolk sac' fry which had just shed eggs but were not yet ready to wriggle out of the gravel and become free-swimming. Many dozen were at the 'alevin' stage which means they had hatched, had finished consuming the yolk sac of food gifted from their mother, and were now ready to swim in the stream as free-living baby salmon. I dug tiny channels in the gravel to give the yet hatched salmon and sac fry a possible path to reach the stream when they fully hatched.

The free-swimming alevins, which had no path to reach the water and were trapped in the dessicating gravel bar, I carefully caught and moved with an aquarium dipnet into the quiet current just below the bar. As insurance, I captured six of the free-swimming alevins, put them in a bucket and brought them to my apartment into a fish tank and kept the water at a constant 65 degrees and bought an extra oxygenation pump because salmon require lots of dissolved oxygen. I kept these six salmon as evidence in case anyone questioned whether these salmon were actually in the stream and being de-watered.

That night, I informed a few dozen local anglers and stream activists of the day's events, and apparently one of my messages was forwarded to the Maine Inland Fisheries & Game department (MDIFW). Using the new power of e-mail I endeavored to create a storm of e-mails to force the State MDIFW to order enough water be allowed down the stream to keep the baby salmon still trapped in the gravel from being killed by lack of water.

I also called Ed Baum, the senior scientist of the Maine Atlantic Salmon Commission and explained the situation and what I had found. Ed said he would do what he could, but said all of his biologists were 150 miles away in far eastern Maine and there was no chance that they could verify what I had observed. He later informed me that the Maine DIFW and the Cobbossee Watershed District had decided against releasing any additional water from the upstream dams to keep the salmon alive because it might affect nesting ducks.

The next evening I had a weird phone call on the message machine at my apartment. It was from a Sergeant at the Maine Wardens' Service who stated he had received information that I was keeping several Atlantic salmon in a fish tank in my house, that this was illegal, and they wanted to talk to me immediately. I knew the laws and knew I was being set up like a bowling pin. Under Maine and federal law, it was perfectly legal for the various dam owners on Cobbossee to dewater the stream and kill all of the Atlantic salmon eggs and yolk sac fry and alevins in their gravel nests. And it was perfectly legal for the Maine DIFW, Maine DMR, Maine Atlantic Salmon Commission and Cobbossee Watershed to stand aside and do nothing and issue them a death sentence. But it was illegal for a person to catch one in an aquarium dip net, put it in bucket and bring it home to a fish tank to use as physical evidence that salmon were being killed. The laws were, and still are, written that way.

So the next morning I brought the aquarium with six one inch salmon in my car, drove back down to Cobbossee and released them into the riffles next to the hundreds and hundreds of dead baby salmon that died in the past two days because the stream levels were not increased by three inches.

In hindsight I have always hated myself for doing this; but frankly I was scared and bewildered. Only nine months earlier I had fired from my job for defending Cobbossee against this gross insult and now I was facing going to jail or being fined $1,000 or more for bringing six tiny baby salmon into my house because I feared they would be killed otherwise. That morning I was to meet with attorneys for Defenders of Wildlife to be a plaintiff on a lawsuit to protect Atlantic salmon in Maine under the Endangered Species Act, but not the wild salmon in Cobbossee, even if we won they would not be protected. But if I were caught harboring these baby salmon my ability to be a plaintiff would be destroyed and the reputation of my new group, Friends of the Kennebec Salmon, would be forever marred by a criminal action that I did actually commit.

So in a fit of cowardice I let the baby salmon go and later denied I ever had them in my possession. And I was too stupid to even take a picture of them in the little tank in my living room, happily swimming around. In the end, me and the salmon completely lost. I lost my only direct physical evidence of their existence and the state agencies were able to maintain the fiction that they never existed and in doing so let the Cobbossee Stream to continue to be de-watered every spring when the dam owners felt like it. I had been used like a tool.

Two Months Earlier

On a cold February morning in 1998 I called up Joseph Emerson, Jr. of Winthrop, Maine, who wore a great grey curlicued moustache and was the long-suffering mill manager of the Yorktowne Paper mill on Cobbosseecontee, owner of the first dam on the stream.

"Mr. Emerson," I said, after reminding him of our previous interviews with the Capital Weekly about the status of the mill. "Does your company use that dam for anything?" He asked why. "Well, I think I can get federal funds to remove it without any cost to the company. If I show that I could, would the company be interested in having it removed?" Joe did a long pause and said, "Well, perhaps. We're still getting our process water from the intake at the dam. But ... we might be interested."

Unknown to me, but as Joe confided a few weeks later, a dam inspector hired by the state had recently visited the dam and gave them a report calling for all kinds of improvements at the dam to make it 'safe.' Actually, the dam was quite safe and I believe consultant was more interested in getting private consulting work for himself than in identifying dams that were ready to fall down. So I carefully pitched to

82

Joe the concept that my offer to remove the dam would get rid of the pesky consultant and save the company the $30-50,000 they were getting ready to spend on an engineering firm to prove the dam was structurally sound. He said he would sound out my idea with his bosses out of state and let me know.

A few months later, Joe gave me the green light but wanted to know details on when and how this removal might happen. He said his deadline was tight because the company needed to respond formally to this state consultant who said the dam was unsafe. By this time I had fallen into the vortex of trying to raise $100,000 from square one to pay for all the costs of removing the dam, from permitting to engineering to removal to site restoration. Many conservation groups in Maine offered verbal support, but after all the meetings, no tangible offers of funding and assistance were forthcoming. I was told the Army Corps of Engineers had a new program called 'Coastal America,' which was designed to use spare Army Corps units and staff to plan, engineer and remove small, old dams on coastal rivers. But the more I learned about this 'program,' after the feel-good powerpoint presentations, the bottom-line was that I or any other group I got onboard was going to have to raise an enormous amount of 'non-federal' matching funds by the way the Corps does everything: expensively.

But we trudged on. Joe Emerson continued to be supportive and helpful, but the mill was still running and the water needed to recycle old cardboard at the mill and turn it into toilet paper tubes was dependent on a small sluice of water that was gravity fed from the top of the dam. Joe said the company had to have assurance they could get water from the stream without the dam before they could formally commit to the project.

But the Corps told me that without the company formally committing to the project they could not even consider putting it into their funding pipeline for the next fiscal year. Catch-22. Other federal agencies told me that while they could put up money to do engineering to remove dam but they could not use this money to figure out how to get water into the mill without the dam. But, I said, that's what the company needs from us for the project to go forward. Sorry, they said, our rules don't allow it. So the project, which everyone agreed was a great project, was stuck in a black hole of regulatory intricasies. And still the salmon and alewives and eels were butting their heads against the dam every year.

A Few Years Later

From around 1991 I had a habit of stopping into the Maine DMR offices in Hallowell, Maine up Winthrop Hill to pepper the fisheries biologists with questions about sturgeon, stripers and alewives and the other fish in the Kennebec. At this time the Edwards Dam was still in place in Augusta at the Kennebec's head of tide and nobody knew when or if it would be ever ripped out or even have a fishway on it. Maine DMR was each spring sucking thousands of alewives from the base of the dam through a giant vacuum cleaner into a tank truck and shipping them upriver to make up for the lack a fishway at the dam.

One day, I asked Jim Stahlnecker, who was in charge of the giant suction device, if they were putting any alewives into the ponds in Cobbossee, above the dams. He said no but they had been talking about it and they still needed to get a final sign off from the local inland fisheries biologists Denny McNeish and Bill Woodward. The next spring, in 1997, Jimmy Stahlnecker said they had gotten their approvals and were going to transport about 2,000 alewives into Pleasant Pond in Richmond and Litchfield from the Kennebec. And they did. They were the first alewives that had swam above the dams in Gardiner on the Cobbossee since the 1760s. We shook hands.

Naturally, I wondered how the alewives would get back downstream and how their babies would get back up. That part was a bit sketchy. Basically, Maine DMR had no plan to get alewives back into the stream except by shipping them up there like frozen peas. This gave me more impetus to work on my scheme with Joe Emerson to get the first dam on Cobbossee removed … and the next two just upstream.

But, I thought, one at a time.

In rainy fall Saturday around 2002 I was up at the American Tissue Dam on Cobbossee, an imposing flat concrete slab of a dam that shoots up 30 feet above your head in a narrow bedrock gorge. It is the only hydroelectric dam on the stream. A five foot slice of water was falling straight over the top of dam which slammed flat onto a concrete step at its bottom and sprayed everywhere. When I climbed down to the dam I saw small fish bouncing everywhere. They were baby alewives. There was no plunge pool at the bottom of the dam.

The baby alewives, coming down from Pleasant Pond were going over the top of the dam and free falling 25 feet and hitting the concrete slab head on. By the hundreds and thousands. I waded into the pools below the dam and started picking them up. Many of their heads were completely crushed. Others had their eyeballs popped out like watermelon seeds, the eyesockets empty. Some had bounced off the concrete so far they were writhing in the tall grass next to the dam. And as I picked them up, hundreds were still coming down over the top of the dam. Downstream, hundreds of them, in pools filled with detached scales tried vainly to swim straight but couldn't. They had severe concussions or fractured skulls. Only the ones that didn't die on impact were noticed by their movement. The rest were pooling up in the glacial potholes of the ledges. It was a Sunday, late afternoon in early October. There was nobody to call. The dam was automated. The State agency staff were all at home. It was getting dark. And I was standing there soaking wet watching thousands of baby alewives have their skulls mercilessly crushed from hitting a stupid two foot concrete slab at the base of a dam.

So I drove down to Reny's department store in downtown Gardiner and bought three of the biggest plastic storage bins they had for $6 apiece and drove back to the dam. I reasoned that if I could use the plastic bins to make a plunge pool of deeper water right below where the alewives were coming over the top of the dam they would hit 2 feet of water instead of a solid concrete. But the force of the water was so strong that they swept the bins away after a few seconds; so I

85

then filled their bottoms with heavy river rocks, but that still didn't work. The force of the water was too strong. So then I just stood below the dam and tried to keep the bins in place with my arms and feet. The river around me was full of thousands of dead and feebly injured baby alewives. And I knew there was nobody to call. So I took a lot of pictures of the carnage and went home and spent the rest of the night sending those photos to every state agency person in Maine. There was nothing else I could do.

At about midnight Sunday, after sending out the photos and descriptions to everyone I could think of, I arrived at a conclusion which made me sick to my stomach. What I saw had been happening every day each fall for the past five years, ever since Maine DMR had started truck transporting adult alewives over the Cobbosseecontee dams in 1997.

It was in part my idea to put them over the dams in 1997. I had assumed Maine DMR would make sure the alewives had safe passage down the stream past the dams. I had assumed they would talk to the dam owners and visit the sites and make sure the route was safe before doing it. I assumed wrong. For the past five years Maine DMR and the dam owners had been annually sentencing these baby alewives to the most cruel and painful death that could be imagined: their eyeballs blown out of their heads and bouncing six feet in the air like popcorn from the concrete slab at the base of the dam.

By this time I had a decade under my belt watching thousands of baby alewives spit into the Kennebec and Sebasticook after being sliced into small pieces by the sharp knives of spinning turbine blades. But this was different. This was like taking them, a thousand at at time, and shooting them out of cannon straight into a brick wall. Their bodies were strewn in the eddies and little stick piles in the stream for a mile.

The next day, a Monday, I was informed the state had told the dam owner to 'fix' the problem; which they tried to do with a plywood box a bit sturdier than my blue plastic bins from Renys, but they still failed. The current falling from the top of the dam was still too strong. A few days later, they put some angle iron reinforcement around the plywood the got the box to hold in the current on the concrete apron. The killing stopped. But it sickened me to think it this had been going on for five years without anybody noticing; and would have never stopped if I hadn't decided on a Sunday to take a walk up Cobbossee just to go fishing.

Not Vacuum Cleaner Hoses

I first noticed large dead American eels in Cobbossee on October 28, 2000 when I was driving back from Topsham, Maine with a backseat full of campaign literature from the printer. I was running for the State Legislature for part of Augusta at the time. I stopped at the stream, since I hadn't been there for awhile, and noticed below the hydro dam a number of long pale white things that looked like vacuum cleaner hoses underwater. I grabbed a long stick and pulled one out. It was an eel, about three feet long, somewhat decayed, without a head or the front of its body. The others in the pool were similar. I reported my discovery the next day to Maine DMR staff who told me they were 'aware' of the dead eels and confirmed they were killed in the hydroelectric dam just upstream. I asked them what they were doing about it. They gave me a blank stare and mentioned about a 'report' to be filed. I went on my way.

The next year I returned to the stream, in the fall, below the hydrodam and saw dead eels everywhere, chopped into pieces, strewn on the stream bottom and draping the hanging branches over the shallows like a sick and twisted person's idea of Christmas garland. I took lots of photos and gave them to Maine DMR. Still there was no response.

In 2002 I went back to the stream in the fall and saw dead eels everywhere and took more time to wade out into the current and count them and again took photos and again alerted Maine DMR. At the urging of two women, Laura Rose Day and Betsy Ham, several of the lower echelon DMR biologists went with us to view the carnage. Myself and Nate Gray spent the afternoon fishing out and counting freshly killed eel carcasses from the half mile of stream below the dam. We came up with about 200 and removed them. Judging by their 'fresh' condition, Nate said most of them had been killed the night before. Nothing was done or said to the dam owner.

Several days later the carnage returned. This time Betsy and Laura got a reporter and photographer from the Kennebec Journal to come down and witness the grisly and counting and clean-up activity. Another 100 or so large eels had been killed. Still, no action was asked of the dam owner. The story made front page of the paper's "B-section."

The third day, after a lot of late night emailing, Maine DMR and the Maine DEP convened a 'meeting' with the operators of the dam, a company called CHI from Andover, Massachusetts, short for Consolidated Hydroelectric. Tipped off to the meeting, I parked along the stream, walked down, saw even more freshly killed dead eels in the stream and then weasled my way into the cluster of officials talking seriously at the top of the dam. The head guy from CHI, John Bogert, told me that for my own safety I should not be walking in the stream below the dam. I told him as a member of the public I owned the stream below the dam and could walk in it whenever I wanted to and did almost every day to pick up dead eels that his dam was killing. He said he needed me to sign an insurance waiver before CHI could 'let me' go into the stream.

So I walked downriver and through the woods and fields to my car, grabbed a wrinkled piece of paper from the back seat, walked back, climbed back onto the dam and wrote an insurance waiver on my knee saying I absolved his company of any claim if I were injured while standing in Cobbosseecontee Stream below the dam. Bogert accepted the paper but noted the company was called CHI now, not Consolidated Hydroelectric, so I changed that and handed it back to him.

A bunch of 'official talk' ensued again so I went back down into the stream to pick up more dead eels. I found one, a female, hung up around a branch near the shore. The back end of her body was decaying from a massive infected gash but she was somehow still alive and her mouth was moving. I ran up the streambed onto the dirt road and up the steep hill to the top of the dam where the 'official' meeting was taking place and stuck the eel in John Bogert's face. "You didn't finish the job, Mr. Bogert," I screamed at him. "This one is still alive but the back end of her body is rotting off." I put the still moving mouth of the eel a few inches from his face while blood from her wound poured down my arm. "Finish the job," I screamed again. "Kill her right here with your own hands." At that point the Maine DEP and Maine DMR officials grabbed me and pulled me back off the abutment, fearing I was going to attack Bogert. When I left, Bogert took his workboot and pushed the dying eel off the face of the dam.

At this time John Bogert insisted that the a 'deep gate' at the dam was being open to six inches, which he said was plenty to let the eels get past the dam without going through the turbines. I had earlier explained to John that if this were true, we shouldn't be finding dead, cut up eel carcasses all over the stream below the turbine outfall. Either the deep gate worked or it didn't. John Perry of MDMR and John Glowa of the Maine DEP then asked Bogert to open the deep gate to 16 inches; and Bogert reluctantly agreed and had the dam operator turn the giant metal wheel which raises the gate, which was under 20 feet of water at the base of the spillway. As the gate was cranked open a giant plug of leaves and sticks and muddy water poured from the deep gate, turning the entire stream below the dam into the color of coffee. After several minutes the water surging out of the deep gate cleared up. They all stood at the top of the dam and watched.

What this experiment showed was that the deep gate had been clogged by a massive plug of leaves and sticks and mud which had backed up and been compressed against the narrow slot in the gate by the suction force of the water. The small amount of flow which had been coming through the gate was being forced through the tiny interstices between the leaves and sticks and mud. There was no way a garden worm, let alone a three foot eel, could have swam through the deep gate. Yet for two decades the dam owner had insisted that the gate was open enough to allow eels to pass through it. Now we had proven they were wrong in a way in which Bogert could not deny because he was standing there.

When John Perry returned to his truck he told me that they were going to test the deep gate at the wide adjustment they had just made for that night and check the stream the next morning for any dead eels. Early the next morning John and I and Nate Gray checked the stream; it was again filled with dead eels. We knew they were from that night because Nate and I had picked up every dead eel from the stream the day before. These had to have been freshly killed; which showed that even with the deep gate open three times wider than before, the eels were still mostly going through the turbines.

When John Glowa was informed of our findings he composed and faxed a formal letter from Maine DEP to CHI's offices in Andover which stated the findings of our one-night experiment: that the wider deep gate opening was still not stopping the eel kill. As a result, he wrote, the Maine DEP was ordering the dam to be shut down from sunset to sunrise for the rest of the eel migration season. For the first fall since it had been rebuilt and brought on-line in 1979, the American Tissue Dam was no longer killing most of the American eels which migrated down Cobbosseecontee Stream. Then there was some more good news.

Do you still want to take down our dam?

Joe Emerson called me out of the blue in 2003. He said the company was shutting down the mill permanently, laying off everyone, but they still wanted to know if I was interested in removing the dam so as to make the property more saleable. I drove over to the mill, on a July day, and sat with Joe in his office.

The mill had been closed for a few months, everyone had been fired, all of the papermaking and saleable equipment sold, transferred or trucked out for salvage. All of the windows of the building were covered with fresh sheets of 4 x 8 plywood to keep local kids from shattering all the windows. The receptionists' office was empty. The whole mill was empty. It was just Joe Emerson coming in every day to supervise stripping the last few pieces of saleable stuff as a way to delay the day when his services were no longer required.

Even though his news made removing the dam much more simpler and do-able it was still very sad. About 60 people in Gardiner had just lost their jobs; many of whom had worked there, doing highly specialized papermaking work, for decades and for which there was no market for their skills left in Maine. The day of the small, specialty paper mill was long gone. Gardiner Paperboard was the last of its kind; and the last functioning factory alongside Cobbossee Stream. It was the end of an era. I felt like a vulture picking over a dessicated, forgotten corpse.

But we trudged on. Joe was right. The parent company still wanted the dam removed, so long as they did not have to pay for it. Not having to supply water to the mill made the engineering for the removal exponentially simpler. Conservation funders were again alerted to the opportunity and responded quickly and with new energy. I embarked on a new and re-invigorated PR campaign for the dam's removal.

By early 2004 basic funding was in place. We went out to bid for an engineering firm to tell us exactly how the dam could be removed, which for many reasons was going to be a delicate and intricate task. This cost $25,000. We were then told by the Maine Historic Preservation Commission the dam was considered eligible for the National Register of Historic Places because of its age and excellent state of preservation. To get them to sign off on its removal would require preparation of a meticulous and elaborate visual and written research project, which even if it met their standards, might still result in them opposing the removal. This would mean that no federal funds could be used to remove the dam, which was bad since about 80 percent of the removal was slated to use federal fisheries restoration funds.

A local land trust had recently acquired the shoreline of the stream opposite the mill, where we needed to bring in construction equipment to remove the dam. We learned some of their board members were not too sure if they actually wanted to let sea-run fish like alewives go upstream. Without the land trust's permission to use their easement, taking the dam down from only one side of the river would double the removal costs and perhaps make it impossible. We also had to rewrite the easement to allow for equipment and to cut some small trees.

We also had the main trunk line of the Gardiner Sewer District running along the stream five feet from the dam and a high voltage power line going from across the stream about 35 five feet over the dam, which would make it impossible to bring in a crane to remove the giant granite blocks which made up the dam. We also had to convince the Gardiner Planning Board, from whom we needed a city construction permit, to let us remove the dam. Some on the board, who had all grown up in Gardiner, were a bit unsure if they wanted these 'foreign' fish like alewives to migrate up the river. But after many weeks of careful coaxing they granted us the needed permits.

By July 2004 with all permits, engineering and funding in hand, we wrote up the bid specifications for a contractor to actually remove the dam and restore the site. Four companies bid and we chose the lowest bidder, Doug Smith, who had recently done two dam removals in Newport, Maine and came highly recommended by state fisheries agencies.

At about this time, my email box was bombarded with requests from federal agencies who wanted to know if we had set a date for the removal celebration and where it would be held and what nearby hotel accommodations were available. I was told a Deputy Secretary of the

U.S. Dept. of the Interior wanted to attend so this itinerary info. was needed asap. I tried to politely inform the federal folks that the dam was still there and there were still 1,000 details that needed to be ironed out before I could even tell the contractor it was okay to come. I said I'd let them know.

The last stitch in the quilt was a set of lengthy negotiations with the Cobbossee Watershed District, whose cooperation we desperately needed to keep stream flows low during the construction work. In these talks we learned that they really didn't support the project and in fact were opposed to restoring alewives to the stream. But because the project was already so far along, they grudgingly agreed to help out. So with this most final of final approvals, I gave the contractor, Doug Smith, a firm date to show up and start pulling out granite blocks: first day after Labor Day, 2004. I waited at the dam at 7 a.m. like a bride at the altar.

But the groom, in the form of Doug Smith, didn't show up. Well, a small bulldozer on a yellow flatbed showed up. But there was no Doug Smith. Me and Matt Bernier, the engineer who designed the dam removal, stood together in the mid morning fog at the shuttered Gardiner Paperboard Mill sipping coffee and looked at the bulldozer on the flatbed like two of the dubs on the cartoon "King of the Hill."

"Well, that's a good sign. But where's Doug?" we both asked. Matt called him on his cellphone. No answer. "That's weird," he said. Another hour passed. Another call. "Hmmm ... maybe he's on his way."

We waited all day. No response. No one showed up. The rest of the week followed this pattern. We called Doug but got no response. Late in the week we got a cryptic message from Doug that he would definitely be there. Then it turned to "might be there." Then it turned to, "I'll do what I can to be there." Then, when we finally cornered him over the weekend he admitted that he had been lying all week and there was no chance of him being there.

By the next week, our construction window had nearly closed. The project should have begun in early August and we had started three weeks too late; by late September the fall rains would make working in the stream impossible, since the Watershed District would have to start dumping water out of the headwater lakes. But these concerns fell into the background when Doug Smith was finally forced to admit he could not do the job. As we only learned later, he owed a substantial amount to equipment rental companies from jobs he had done that summer, and none of them would rent him the excavators he needed to do our job until he paid them. As a last ditch effort, I offered to front him the money to put a deposit on the equipment, since we had the funds on hand to do the job. Apparently, his outstanding debts were beyond the $5,000 in free cash we had left in the project bank account. For me, as I stood next to dam, in all its 'still there-ness,' watching it has I had for a decade, it was like I had just woken up and all this 'construction' stuff was like the weird dreams you have when you have a fever.

Izzy Feldmus to the rescue

In spring 2005 we learned that the mill owner, Newark Paper Group, had sold the dam and mill property that winter without telling us. I wasn't too shocked since we had shown them in the fall of 2004 we had failed to do what we had told them we would do: remove the jeezly dam. The new owner was a burly, garrulous Jewish man in his 60s from Brooklyn, New York who looked and talked like he thought Harley-Davidsons were for effete suburbanite poseurs. His name was Israel Feldmus, but he insisted we all call him "Izzy."

We all met Izzy at a law office in Augusta with his son, Aaron, who was in his late 20s. Izzy bought the whole shebang for a song and as we later learned, specialized in buying, refurbishing and re-selling old industrial sites like the Gardiner paperboard mill. He knew nothing about fish, nothing about rivers and nothing about alewives, but the more we yakked and then went down to the stream to the mill, he and I made some type of pre-verbal connection. Although he was a very rich man, he kind of dressed like a hobo, but not as an affectation. It was

94

more that it seemed like he had too much other important stuff to do --
deals to make -- to worry about how he looked. It probably helped him
in negotiations with slick real estate lawyers: the way he presented
himself, they took him to be rube while he ended up taking them.

By June of 2005 we were back on-line. Izzy gave us the green
light to hire a demolition contractor and get the job done. His only
request was that we had adequate liability insurance coverage. Having
been stung the year before by Doug Smith, I immediately called a very
large demolition company in South Middleborough, Mass. called
Costello Dismantling. They had been a bidder the year before but were
a bit high. Richard Walsh, their bid manager, told us they would honor
their previous year's bid of $100,000 for the job and said they could do
it as soon as water levels in the stream went down in late July. So we
were on. Woo hoo !!!

Then I got a short email from Jeff Reardon of Maine Trout
Unlimited. It said, "Izzy Feldmus died over the weekend of a heart
attack. Aaron sent us a phone message."

Izzy didn't leave a will. Izzy had lots of children and relatives.
Izzy's real estate holdings were extensive and complex, stretching all
over the Northeast. His estate was going into probate. So much for
2005.

Many months later, the Feldmus family sold the Gardiner mill
site and dam to a guy from Winthrop who, after being approached,
stated he wanted to keep the dam. In 2009 the mill site was nearly
burned to the ground by local teenagers, several of whom were arrested
and charged and convicted of arson. The dam, being made of soaking
wet granite blocks, did not catch fire.

In 2010 the almost completely destroyed property was bought
for a song by a private group of investors based in Freeport. I talked to
one of their principals in the summer of 2010. He was quite nice and
said various agencies had already approached him about his interest in

removing the dam. He said he was intrigued but first wanted to explore the possibility of reconstructing the dam into a working hydropower project. First he said, he had to deal with the Maine DEP about all of the toxic waste that was still believed to be on the site.

Chuck Loon: Man with a Mission

After the debacles of 2004 and 2005 I took a personal hiatus from working for fish passage at Cobbosseecontee Stream. Thanks to Lewis "Chuck" Loon, who became the dam operator at American Tissue after a change in ownership in 2004, the turbine intake was equipped with a steel punch-plate screen which completely prevents eels from getting into the turbines. Chuck invited me to personally survey the stream with him after its installation and we both confirmed by wading up and down it that the eels were no longer going through the turbines and were using the wide open deep gate to pass downstream without harm.

Chuck is a local guy in his early 40s who grew up in the town of Richmond, just down the Kennebec from Gardiner. In the early 2000s he worked the 7-3 shift at Bath Iron Works and to make some extra money did part-time work at the American Tissue Dam when he got out of work at BIW. Chuck was the guy who showed up at the dam in the late afternoon to make adjustments to the turbines and found the dead eel carcasses that I would stack at the front door of the powerhouse, hang from the door knob, and drape over the sign at the top of the driveway. Since I would not go down there when I saw a truck at the powerhouse, for fear of getting kicked out, we had never met.

Our first introduction was the day that I stuck the bleeding eel in John Bogert's face. Chuck was the guy who turned the big iron wheel to raise the deep gate at John Perry's request and then found it was completely clogged with leaves and mud. It turned out that Chuck had seen the story in the Kennebec Journal about the massive 2001 eel kill.

As he later told me, the story had a profound effect on him, in part because his friends and family also read the story and asked him what the hell was going on at the dam. "I have a small daughter," he said in 2004, as we rode in his truck. "And what was going on at the dam bothered me, because I worked for them. I saw the dead eels in the stream just like you did, but CHI wasn't interested in doing anything to stop it." Chuck told me that as early as 2002 he had proposed to CHI his idea of screening off the turbine intake to keep the eels out of the turbines. "I asked them to at least let me try it and see if it would work, but they said no."

In late 2003, the American Tissue Dam was sold to a New Jersey company, Ridgewood Power, led by a man named Charlie Weymss. His first act upon taking ownership was to fire CHI as the dam manager/operator, but he asked Chuck if he was willing to stay onboard and work for him full-time. One of Chuck's first acts was to ask Charlie if he could try his idea of screening off the turbine at American Tissue with steel punch-plate, the idea that CHI had rejected. Charlie gave him the green light and in September 2004 the screen was designed, built and installed at Chuck's direction. Not only did it completely stop the eel kills, but it allowed the company to get out of having to shut the turbines down from sunset to sunrise every day in the fall as they had had to do since the 2002 eel kill debacle. Chuck also discovered that, within a few percent, the punch plate screen did not affect the ability of the dam to generate electricity.

One October day in 2004 Chuck stopped at my apartment building in Augusta and knocked on the door. I didn't know who he was. He came in and introduced himself and said, "You want to take a ride down to Damariscotta? I want to tell you some stuff."

As can be seen from the all of the above, the problem of fish kills at dams makes it difficult, if not impossible for people like me and people in Chuck's position to actually sit around and shoot the bull. By this time I had become quite well known and a bit notorious in the small Maine world of hydro dam owners and operators.

From the perspective of the hydro industry, Chuck was risking his job by just coming over to my apartment because I had been raising high holy hell in the fall of 2004 about a massive eel kill at the Benton Falls Dam and a large kill of baby alewives at the Burnham Dam, which was owned by Chuck's boss, Charlie Wemyss, who had a few days earlier intimated I could be arrested as a suspected terrorist if I again went to the dam unannounced on a Sunday morning to photograph chopped up baby alewives.

So when Chuck Loon showed up at my apartment at 3 in the afternoon and wasn't there to serve a court order on me, threaten me or beat me up, I figured it would be a good idea to listen to what he had to say. So off we went down the stairs, into his company truck and downriver to Damariscotta, talking the whole way. It was in the truck that Chuck told me his story; he already knew much of mine.

For Chuck, the whole problem of getting fish past dams without killing them was just like any other problem that needed to be solved: you sat down and tried to figure out how to solve it. What made Chuck unique was that he considered the problem worthy of solving and was sure there had to be a solution if you looked hard enough. Up until my arrival at Cobbossee, he said, there was no problem, since it was considered okay to kill fish so long as some agency didn't tell you to stop. Basically, Chuck said, CHI had decided it was cheaper and easier to deny there was a problem than fix it. "That made no sense to me," he said. "Because I saw the dead fish just like you did, and it bothered me that we were doing it, so I figured there had to be a way to stop it, and if we stopped, you wouldn't be putting dead eels on the door knob when I showed up in the afternoon." He laughed and said, "That was me, you know, who found all those dead eels you left."

As we rode through the old farms and fields of Whitefield, over the Sheepscot River towards Damariscotta, I kept waiting for Chuck to begin lecturing me about how I had to be 'reasonable.' But he didn't. He was a man on a mission, high on the belief that there was a practical way for people like himself and myself to find a sense of harmony that didn't involve either side having to give up anything. Mostly, I was struck by how openly he admitted that my concerns were legitimate: that it was wrong for dams to kill fish. "But for us to figure how to fix it, people like me have to be working work with people like you," he said. "If all we do is fight and argue, nothing ever gets done." Wow. No dam owner or operator had ever spoken like this. But I still kept waiting for the other shoe to drop. Then Chuck burst into another story.

"Last week I was at a meeting with all of the other dams on the Kennebec and Sebasticook," he said. "And we were all talking about eels, because of what's going on at Benton. It drove me crazy," he said. "Bob Richter [a fisheries biologist with FPL Energy] was going on and on about how they needed to do all kinds of studies to see if they were killing eels, and if it was a significant amount and what could be done," Chuck said. "And I said, 'Bob, it's really friggin' simple. If you look below your dam and see a hundred dead eels all over the bottom, you know you've got a problem !!!'" By this time Chuck was almost screaming in exasperation.

As he talked I discovered that Chuck was by nature a fixer, a handyman: a mechanically skilled and intuitive guy who was as good with a tool box as I with written invective. Chuck viewed the eel kill at Cobbossee the way you'd view a carburetor that was out of adjustment. It was just a problem that needed to be fixed and had a rational solution. So you went to work and tried to fix it.

Chuck viewed the problem at hand as how to keep the eels from getting killed the turbines. The most logical way to do that was to keep them from getting in the turbines in the first place and giving them another place to get past the dam. So Chuck's idea at American Tissue was to buy a large sturdy steel plate filled with thousands of half inch holes and bolt it onto the upstream face of the turbine intake, 15 feet underwater. If the holes are the right size and at the right spacing, he said, they are too small for eels to through but big enough to get water through. Water goes through; eels don't.

"But I didn't know what effect the screen would have on the turbines, flows and generation," he continued. "But the only way to tell was to put it on and see what happened and make adjustments. The whole thing costed out to about $2,000 and Charlie said to go ahead, so we did. That's what we were doing in late September when we drew the dam down. We were installing the punchplate."

To make sure that eels weren't being impinged and pinned against the screen by the suction of the water, Chuck told me that he had invited Nate Gray and Skip Zinck of Maine DMR to come up to the dam with an underwater video camera attached to a long pole. "They looked with the camera along the screen and we didn't see any eels pinned on it. I think what's happening is that the because the deep gate is close to the intake and we have it pretty wide open, they end up going to the deep gate."

We got to Damariscotta and parked at the dam powerhouse. Damariscotta Lake is a long natural lake with an outlet stream that drops about 100 feet in a quarter mile into a narrow saltwater estuary called the Damariscotta River. The hydro facility works by sending water down from the lake in a long steel tube about 4 feet in diameter to a small powerhouse built in the 1920s. Chuck took me inside and described how the entire place had been in disrepair when his company and bought it and all of the improvements he had made and still needed to make. It was very obvious that he loved a job where there were lots of things that needed fixing or could be somehow made to work better. Chuck was not a clock watcher.

He explained that Damariscotta was an anomaly in Maine. In recent years, he related, the previous dam owner had voluntarily decided to shut down the turbines in the fall because of the difficulty in keeping juvenile alewives and eels out of the penstock and going through the turbines. As a result, the dam was perhaps the only hydro dam on a coastal river in New England that didn't kill fish. When we were there the turbines weren't generating, per the usual fall custom, and Chuck described some ideas he was toying with about using his experiments at Cobbossee to see if they could start doing some fall generation by employing screens to ensure no fish could get into the penstock.

"Now at Burnham," he said, referring to the dam he managed on the Sebasticook River 60 miles to the north, "We're looking at screening off that penstock the same way we're doing at Cobbossee and that will be part of the permanent downstream passage that we need to get in next year."

Then we started driving back to Augusta. As we crossed the Kennebec River at the Randolph-Gardiner bridge, Chuck suggested we go to Cobbosseecontee to look for eels, adding that he had a spare pair of chest waders I could use. I said, sure, let's go. As the sun started to go down we drove down to the powerhouse at American Tissue, suited up in waders and started walking up, through and across the turbulent, slippery rapids where I had spent days and weeks in the past few years pulling out dead, chopped up eels. Chuck never stopped talking once, like a guy who had been in solitary confinement and just released to blue sky and fresh air.

After about an hour of careful looking, we couldn't find a single dead eel. Chuck even suggested looking to see if any eels had gone over the top of the spillway and had been killed. There were none up there either. And no dead baby alewives. I said, "Well, Chuck, this is exactly the time of year when we would be seeing dead eels here if they were here. I know every single rock in this place where they tend to bunch up, and there aren't any."

"So are you convinced?" he asked, as we got out of the stream, sat on the tailgate and pulled off our waders. "I have to say yeah," I replied. "We've given the place a good scouring and there aren't any." He replied, "Good. I wanted you and me to do check it ourselves so there was no question."

When we drove up the river to Augusta so Chuck could drop me

101

off, we ended up back in my apartment because he was still talking. "Here's something I want you to think about," he said, "But you don't have to give me an answer now. I want you to think about working for us, as a kind of a consultant, so you can help us figure out how to make these dams work right for the fish, since you know so much about it. I haven't talked to Charlie about it, but we'd pay you. Because the way I see it, if you say the thing works for fish, then I know it does."

"Wow … " I said. "I'm not sure if Charlie Weymss would really want me involved like that, especially if he's paying me."

In the end, Chuck's idea of hiring me didn't come to fruition, mostly likely because his boss put the kaibosh on it as being too strange an arrangement to even consider, but I don't know for sure and never asked Chuck for more details. Frankly, I had deep reservations as well, which I didn't share with Chuck, because I sincerely believe there are still too many dams on our coastal rivers and more need to be removed. It would have been awkward at best for me to be getting a check from American Tissue Dam while at the same time scheming ways to get it ripped out of the river. But I cannot deny the straightforward logic of his approach.

What most struck me about Chuck Loon was he was a guy who worked a full shift building navy ships, took a small part-time job at a local dam to earn some extra money, and in the span of a few days or hours mulling it over, independently devised, designed and installed two ingenious and highly effective downstream passage systems for alewives and for eels. All's he needed was a boss who let him do it. And it works. Chuck accomplished by himself in a couple weeks what the entire global cadre of engineers and scientists have yet to be to do – to make a dam so it doesn't kill fishing swimming past it. In the end, Chuck succeeded because he viewed the problem as real, as solvable, and wouldn't take no for an answer.

Now we just have to clone him.

Sisyphean's just a word for nothing left to lose.

It's very unclear if fish passage will be restored to Cobbosseecontee Stream in my lifetime; having now spent more than a decade trying to get it with very little so far to show. It seems the human obstacles to getting fish passage at this stream are as massive as the dams are to the fish themselves. Absent a willing owner at the first dam, the old Gardiner Paperboard dam, removal is off the table. The State fish and game commissioners could use their legal authority to order fish passage at the dam but have shown no interest in doing so; and pursuant to *Dumont v. Speers* citizens cannot compel the Commissioners to undertake a proceeding to determine if fish passage should be constructed at the dam. It literally depends on the whim of the appointed Commissioners themselves and the support of their biological staff. If the citizens and city officials of Gardiner were strident and vocal in enough in their support for fish passage, the Commissioners would be more likely to take action, but this support has not yet developed.

Chain Pickerel:
The First Alewife Substitute

Many of the fish in the rivers and ponds of New England did not live here 200 years ago. They were put here by people. And most of the fish that did live in New England 200 years ago do not live in them today. Dams and pollution extirpated them. It's as if our chickadees, robins, cardinals, squirrels and deer and moose were replaced by kangaroos, Galapagos finches, gazelles, monkeys and tortoises and nobody noticed.

The chain pickerel (*Esox niger*), called by fishing writer A.J. McClane, 'the wacky king of the weed empire,' is native to coastal Massachusetts but not to most of Maine. It was the first fish ever deliberately 'transplanted' in Maine.

Today the pickerel is not, as they say, esteemed as a food fish. It rarely grows large, is as thin and snaky as an eel, and what scant flesh it has is rife with tiny, "Y" bones that have the nasty habit of sticking in your throat when you swallow.

But in the early 1800s, Maine underwent a pickerel stocking craze, as settlers transferred them in a daisy chain from one pond to the next, not for fishing, but as food to eat. The roots of this stocking craze are not totally apparent, but a few facts come to light. The first is that two of the earliest pickerel stocking sites were watersheds which had lost their alewife runs due to impassable mill dams, Cobbosseecontee Stream in the lower Kennebec River and Davis Pond in the lower Penobscot River.

Samuel Boardman of Augusta stated in 1864:

"I was, many years ago, told by some of the old settlers on the Kennebec, that there were no pickerel ever seen in the tributaries on the west side of that river, in the vicinity of Augusta, until they were transferred there from Togus lake, which is on the east side, by the late Robert H. Gardiner. They were put into the Cobbosseecontee Stream above the mills in Gardiner. The few put in at that time have multiplied and spread into all the connecting waters, and are now, with the exception of the perch, the most abundant of any fish species found there."

This is a very interesting recollection since Robert H. Gardiner, whose family owned the impassable mill dams at Gardiner on Cobbosseecontee Stream, fought for many decades in the late 1700s to not put a fishway at his mill dams to allow alewife passage. Gardiner finally won the battle in 1806 when he convinced the Massachusetts Legislature to exempt Cobbosseecontee Stream from the state's fishway laws. Apparently, although this can only be guessed, Gardiner's decision to physically stock chain pickerel above his Gardiner mill dams might have been intended as a mollification to upriver settlers, ie. to give them a fish to eat that did not require passage at his dams, and thereby to get them to stop appointing 'fish committees' at town meeting each spring to force him to build fishways for alewives.

In any event, the existing records show that the practice of stocking chain pickerel became a *cause celebre* in Maine during the first half of the 1800s. In their 1869 Fisheries Commissioners Report, Atkins & Foster state at p. 90:

"The fact of pickerel displacing the trout when the two are brought into contact is too well known to need a demonstration; but we will cite some facts to show the extent to which this has occurred. 'All of the ponds and streams,' says Mr. C.T. Chase of Dixfield, 'which empty into the Androscoggin from the north above the outlet of Wayne pond were originally stocked with trout exclusively, excepting the small fry, such as minnows, smelts, chubs, shiners, etc. The streams emptying in from the south were stocked mostly with pickerel, catfish, brindle perch, with the small fry named above. Many of the ponds and streams emptying in from the north have been supplied with pickerel; and in some instances the trout have entirely disappeared, in others they seem to partially hold possession.'

"On the Kennebec the waters on the west side were stocked with trout and no pickerel, and those on the east side below Skowhegan with pickerel. Further north, we find that pickerel were brought from Madison pond into the waters of Gilman stream, a tributary of the Carrabassett. Twelve or 14 years ago pickerel were introduced to Dead river. At that time trout were very abundant, but now have almost entirely disappeared from that portion between Chain lakes and Grand falls, while the pickerel have become numerous. There are said to be pickerel in Pierce pond, but no farther north in the Kennebec.

"Williamson says: 'This species of fish was first brought to Penobscot County in 1819, and put into Davis pond in Eddington, where they have increased surprisingly; but they devour the white perch, which is of as much or more value, and their immigration has not received much welcome' Pickerel were unknown to the upper Penobscot previous to 1824. That year they were introduced into Mattanawcook dead water in Lincoln, and they have now reached Seboois Grand Lake, at the head of Seboois river; on the west branch they have not yet reached Millinocket and Pamedumcook lakes. From the Penobscot they were transferred to the St. Croix twelve or fourteen years ago, by carring a number of them forth from Baskahegan lake to Little Musquash lake. From the latter lake they have spread into all the St. Croix waters; the trout (Salmo fontinalis) has given place to them on the lower lakes, but the Schoodic salmon and togue stand their ground better. From the St. Croix they have been transferred to Meddybemps lake, on the Denny's river, and have in some way obtained a footing in the East Machias River, where they have become abundant.

"It is worthy of note that in all these cases the pickerel have increased rapidly. Dr. Cochrane says the eight pickerel were put into Cochnewagan pond in 1825, and in 1833 they had increased to such an extent that he was able to catch a handsome string of them, averaging three pounds in weight. And it is worth while to consider whether our ponds do not yield a greater weight of pickerel per acre than they would of trout. The quality of the former is generally considered to be inferior, but we are not aware that it is less nutritious.

"We advise that legislation should forbid the introduction of pickerel into any waters where they do not now exist. The same prohibition should rest against sunfish and yellow perch, and the indiscriminate introduction of black bass should not be permitted.

"As between white perch and pickerel we would make a distinction in favor of the former, which is one of our best fishes, lacking only in size; but the extent of the injury done by the pickerel is not sufficiently clear; they seem to flourish well in the same waters. Anebescook lake is, at the same time, the most productive of pickerel and white perch of all the lakes of that system. Yet it is probable that the perch would be more numerous without their persecutors. We think some careful observation is necessary to determine whether more food is produced from a lake stocked with white perch alone or from one stocked with white perch and pickerel together. Until we can answer this question, it is doubtful whether one of these should be protected at the expense of the other. We therefore refrain from recommending any other legislation in reference to pickerel than that they be not introduced to any new waters. For the protection of white perch we suggest that none should be taken during April and May, except with the hook; that would cover their spawning time*; but perhaps that action is not necessary." (Fn. in original at *: "Incorrect: they spawn later.")

It's easy today to track the dates of each pickerel stocking in Maine in the 1820-1850 period, because within a year or two after a pickerel introduction, a petition or law would appear in the records of the Maine Legislature by local residents protesting the over-fishing of pickerel in that pond and demanding restrictions on the size of the catch. From 1800 to the Civil War, more petitions were submitted and laws were passed in Maine regarding chain pickerel than any other fish. Why the plethora of pickerel planting and protective pickerel petitions?

It's not hard to review these legislative petitions and acts in Maine from the late 1700s to the Civil War and not see the curious intersection of alewife and pickerel protection laws in Maine. In fact, a recognizable pattern quickly emerges. As mill dams systematically wiped out native alewife runs and laws requiring passage for them failed to be enforced or were repealed, pickerel were manually stocked in the former alewife ponds to replace the alewife as a source of sustenance to the early settlers. Then, rampant pickerel overfishing commenced, and a new round of petitions and laws were passed to protect the pickerel, since the alewives were long gone and the pickerel were all that was left.

Like the earliest Biblical apocrypha, which could only be read by reading a 'palimpsest,' meaning scrutinizing the faint words in a tanned goatskin underneath the words later written on top of them, we have to turn to mid 19th century histories.

Late 19th century histories of the early settlement of towns in Maine and Massachusetts contain a lot of flowery prose of dubious authenticity, except in those instances when the authors quote from actual documents from the time of settlement. Much in them is made of the ardor and starvation and privation of the early settlers; how they had to hack down dense forests of six foot wide white pine and oak with pen knives just to scatter a few moldy corn seeds from a leaky bateau and then watch as the crows ate them or a late snow killed the seedlings and they were forced to subsist by sucking out the faint food-like molecules encrusted in rotted burlap, salted mutton, all while praising and thanking Divine Providence.

Most of these late 19th century historians were well-to-do lawyers or preachers who either self-published their works or were hired by the townspeople to collate the town's history for a centennial celebration, and in doing so, understood it best for book sales to discretely leave out the bad and tell only the good, at least about those whose families were well-to-do and still in town.

To read much of J.W. Hanson's text in his History of Gardiner and Pittston (1852), Maine one would think the settlers of the Gardiner area in the 1700s scraped by eating tree bark, army worms, winter-killed wheat stems and the leather off their highly worn Bibles, but then, Hanson, without comment, drops this bombshell, from an unpublished manuscript of Gen. Henry Dearborn (b. 1783) who grew up in Gardiner two decades after it was first settled:

"Major Seth Gay built the first wharf and Gen. Dearborn established the first ferry (across the Kennebec) in 1786. He was accustomed, as were others, to draw a seine around the mouth of Cobbossee Contee, and incredible numbers of shad, herring, salmon and sturgeon were taken every spring."

Forty pages prior, Hanson notes that, "Jonathan Winslow used to relate that he captured 16 noble salmon one Sunday before breakfast."

Now let's remember that a 'noble' Kennebec salmon at the time ranged from 15 to 30 pounds. So, on just one Sunday morning, before Church and breakfast, Jonathan Winslow procured (most likely in a seine) nearly 300 pounds of fresh Atlantic salmon. The utter privation !!!

And if you were a housewife tired of eating smoked alewives in the 1770s, you could always trade them for furniture or a spinning wheel. Hanson writes, this time of Worromontogus Stream, across from the river from Gardiner, Maine:

"It is related that alewives were so plentiful there at the time the country was settled, that bears, and later swine, fed on them in the water. They were crowded ashore by the thousands. Mrs. David Philbrook, who was a McCausland, was very much in want of a spinning wheel. One day she took a dip net, and caught seven barrels of alewives in the Togus, and took two barrels in a canoe, and paddled them down to Mr. Winslow's, and exchanged them for a wheel."

We'll remember here that the Mr. Winslow who sold Mrs. Philbrook a spinning wheel for two barrels of alewives was the same Mr. Winslow who caught 16 'noble' salmon before breakfast on the Sabbath. Apparently Jonathan Winslow was a fish monger and a trader.

The Rev. Paul Coffin toured the Sebasticook River in July 1796 and reported in his diary:

"July 30th, Clinton. Rode two miles to Capt. Jonathan Philbrick's on Sebasticook, just above the falls, where they catch herring and shad. Thousands of barrels of herring have been taken this spring. They put four quarts of salt to a barrel of them, and when salted enough, they smoke them. They are then handy and quite palatable. Mr. Hudson had three thousand of them hanging over one's head in his shop or smoke house. A pretty sight."

Author Sanger Mills Cook provides a unique glimpse at the first white settler of the land at the junction of the East and West Branch Sebasticook River, Lovel Fairbrother:

"Lovel Fairbrother came to the Kennebec at an early day and explored this river and the Sebasticook; found choice intervale at or near the fork of the river, and abundance of fish in the river and game in the forest. He therefore pitched his tent a big camp near the forks of the river in 1775 and moved his family there being joined by two others and this commenced the settlement in what is now the prosperous town of Pittsfield, then called Sebasticook.

"Soon after he got his family there, he was visited by the Plymouth Patent surveyor, who was surprised to find a man of his intelligence in that secluded place to which there was no road; separated from all other settlements by ponds and swamps and impenetrable forests and he took from his haversack a bottle of rum and instated him as Governor of Sebasticook and treated him and he was then called Governor as long as he lived. [8]

"The Governor was disappointed in his expectations. He did not enjoy living upon herring and coarse bread made of pounded corn. There being no mills within 20 miles and no road or communication with other places but by water in the summer and ice in the winter. The land being on Plymouth Patent he could get no title to it; and could have a deed of a lot given to him if would settle in Norridgewock. He in 1777 transferred his possession at that place to Moses Martin who moved there from Norridgewock with his family and spent his days there to old age."

Samuel Boardman's 1864 report to the Maine Legislature gives an insightful account into this period. First, he provides a unique glimpse of sustenance salmon fishing on the Kennebec River in Vassalboro in the early 1800s:

"An aged woman, who formerly lived on the banks of the Kennebec in Vassalboro, and who, at that time, had a large family of children to support, once told me that, in spring and early summer, the fish from the river were a very essential aid to them -- that many times she has sent one of her boys down to the river early in the morning to

---

8   The patent surveyor was likely Ephraim Ballard, husband of the nurse and
    midwife Martha Moore Ballard of Augusta, who achieved posthumous fame in
    Laurel Thacher Ulrich's 1990 Pulitzer Prize winning historic study, *A Midwife's Tale*.
    Moses Martin is most likely the namesake of Martin Stream, which enters the East
    Branch Sebasticook several miles below the outlet of Sebasticook Lake. Martin
    Stream's source is Plymouth Pond, named for the Plymouth Proprietors.

catch a salmon for breakfast, with as much certainty that he would bring one home in season, as if she had sent him with the money to a city fish market, where she knew they were kept for sale."

Next, writing about *his* present day, with most of the state's sea-run fisheries almost gone, Boardman sums up the crisis for the Maine Legislature:

"Everyone now knows that salmon, shad and alewives, and indeed all the other kinds of migratory fishes -- those that spend winters in the salt water, and come up out of the sea at certain periods, as if sent by a kind Providence, to spend the spring and summer in fresh water -- are now very scarce indeed, and in some streams totally extinct. Everyone knows, too, that many of the species of fishes which remain permanently in our fresh waters, have very much decreased in numbers, as well as in size and fatness. People say that this is a necessary consequence of the building of dams and mills, and filling the streams with obstructions of various kinds for the industrial pursuits of a civilized community. No doubt it is a consequence of these obstructions, but it not need be a *necessary* consequence. I hold that dams and mills might be constructed, and continued, and yet by a little concession on the part of dam and mill proprietors, and a more general diffusion of the knowledge of the natural history fishes, more intimate acquaintance with their peculiar habits, instincts, and wants of life, the mills might remain and the fish continue to perform their annual pilgrimage to and from their breeding haunts, if not in so great numbers as in former times, yet in such numbers as to afford a vast amount of provisions and even luxury to the communities which are now wholly deprived of them."

"I am also aware that this subject has been discussed over and over again -- that for years and years past, every session of our Legislature was thronged, and committees were worried and teased by mill owners on the one hand and fishermen on the other -- one demanding the privilege of building dams and mills without let or hindrance as to the fish, and the other pleading for some reserve, some fish-way, or some accommodation to the annual flow of the fish, which had been of such signal service to the support of the people on the banks and vicinity of the waters in question. I am also aware that our Legislators, actuated by a sincere desire to do justice to all parties, and to give equal rights to all, have, in most instance, made provisions in the several charters and private acts pertaining to mill owners, for the passage of fish at certain times and seasons, with a hope that, while it encouraged the establishment of mills and machinery, there would be also at the required times a safe and successful transit for the various species of fishes that required such passes as one of the indispensable requirements for the continuation of their existence. And we are all aware also that, either from ignorance of what habits of the fish demand, these ways have not always been properly constructed, or from selfishness in mill owners in not keeping them open at suitable times, these provisions in most cases failed, and the destruction of the fish is the inevitable result. "

What's important to note in Boardman's statements is that he asserts the importance of native, sea-run fish as a source of domestic food and sustenance for poor families, as opposed to citing their value as a fungible product for export. This is interesting because we know from several informants that by the very early 1800s, the catching of sea-run fish had been fully industrialized on the large rivers of Maine, especially the Kennebec and Penobscot, using capital provided by investors in Boston and New York and employing local, seasonal workers. These operations, in conjunction with mill dams, made it nearly impossible for a poor family to continue using sea-fish as a source of daily or weekly food. The fish were no longer abundant enough to be caught without fairly specialized and intense effort, especially upriver, well above the head of tide.

This is evidenced by author Carleton Fisher's description of the alewife and shad fishery in the Sebasticook River, where he reveals that the bulk of the alewives caught were immediately purchased, hard-salted and shipped to the West Indies to feed slaves in slave plantations.

"George Sullivan Heald described the fishing activities of his father, Capt. Timothy Heald. Captain Heald was living on the Sebasticook in Winslow, but his activities will give some indication of the fishing industry in the area. During 1797 he had a fish seine catching shad and alewives, for which he received one thousand dollars besides some material for building a house. The fish were transported to market in a large box made by laying a double floor of boards twenty feet square, placing boards around the outside until it would hold forty barrels, then the top was covered with two thicknesses and the corners bound. These fish were sold for one dollar per barrel and sent to the West Indies for the Negroes."

This is also stated by John Godfrey in his 1882 book, "Annals of Bangor," describing the city's waterfront on the Penobscot River in the 1790s:

"The fishing season, in the spring, continued about five weeks; time of greatest plenty, two weeks. Salmon were taken during three months at least, but they were not abundant. From $1 to $1.25 per barrel were paid from the vessels for alewives, and what were then considered fair prices for shad. Newburyport vessels were engaged in the trade and took large quantities of fish to the Southern markets and the West Indies for plantation purposes."

In the same volume, Bangor resident Joseph Carr describes what the boats loading up with Penobscot River alewives were off-loading:

"In the year 1806 my father built a wooden store now standing on Washington Street at the City Point, between the brick stores built by Zadoc French and Joseph Leavitt, and the wharf known as 'Carr's wharf,' which was the first wharf built on the Penobscot River. In this store my father traded until about the year 1842. All sorts of goods were kept for sale, and Saturday was the great day of trade, and Saturday afternoon (my just holiday) was usually spent by me on compulsion in waiting on my father's customers. On this day there came to the store men from celebrated families of Harthorns, McPhetres, Spencers and

114

Inmans, bringing with them shingles, salmon, shad, smoked alewives and credit, for which they wanted tea, tobacco, calico and rum. It was one if not my chief duty to quench the thirst of these most thirsty customers. Innumberable gills, pints and quarts of good old 'Santa Cruz' have I drawn and delivered to these genial souls, of whom I can truly say none were drunk, but 'all had a drappie in their' ee.' I have now in my possession the original copper gill cup, which furnished those hardy pioneers what they considered to be almost their 'meat and clothing' and their drink it certainly was."

The 'Santa Cruz' rum which was off-loaded at Carr's Wharf in Bangor was made from sugar cane in the West Indies, grown at the islands' enormous African slave plantations. Each spring, the ships loaded up with heavily salted alewives to be sold as slave food to the West Indies, whereupon the ships traded the alewives for rum which was then brought back to Maine to be sold to the local populace. This severely under-reported fact shows how closely Maine was intertwined with the global slave economy of the late 1700s and 1800s, and how Maine and New England alewives were a key natural resource export in the 'triangle' slave trade prior to the U.S. Civil War.

The closure of this period of abundance came when the sea-run fish were so heavily exploited that they were no longer numerous enough to be viable as an export product. This occurred first due to massive over-fishing, and finally by the completion of large dams at the head of tide of all of Maine's larger rivers in the 1820s and 1830s, which was the death knell of the fish themselves.

A May 1829 news item in the *Bangor Register*, as reprinted in the Augusta, Maine-based *Kennebec Journal*, states: "A true fish story -- Seven thousand shad and nearly a hundred barrels of alewives were taken in Eddington last week by Luther Eaton, Esq. at one haul -- Bangor Register."

Thirty years later, Maine Fisheries Commissioners Charles Atkins and Nathan Foster stated in 1869 of the Penobscot:

"During all these early years the fish found extensive breeding grounds above the occupied portion of the Penobscot valley. Though shut out from some of its tributaries, a circumstance alone sufficient to effect, in time, a decrease in their numbers, the great highway to the many lakes and streams in the wild lands remained open until about the year 1830. It was then nearly closed by Fiske and Bridge's dam at Oldtown Falls, in which there was and still is a passage by which some salmon pass every year; and in favorable seasons shad and alewives pass in limited numbers. After this the Great Works dam was built, and in 1834 or 1835 the Veazie Dam. The latter was closed in the winter. When the fish came in the spring they found an impassable barrier across their way; they gathered in multitudes below the dam and strove in vain to surmount it; many returned down the river, and after the usual time for spawning of shad was past they were taken in weirs in the town of Bucksport, loaded with ripe spawn they could no longer contain; a phenomenon which Mr. John C. Homer who has fished with weirs at that point for forty-three years had never observed at any other time. These were doubtless shad whose natural spawning grounds lay far up the river, and who had after long contention given up the attempt to pass the Veazie Dam. A great many shad and alewives lingered about the dam and died there, until the air was loaded with the stench."

These documents show sea-run fish were wiped out on Maine's rivers, especially its largest and most productive, by three forces which worked synergistically. First, the fish were caught in such massive numbers that their populations quickly plummeted. Second, as the populations grew smaller and smaller due to over-fishing, highly capitalized corporations moved in to erect large dams on Maine's largest rivers, arguing that there were so few fish left that nothing of value would be lost if the last few were wiped out. Third, the numerous laws passed to control over-fishing and dams were so poorly enforced to be ineffectual.

A major question here is whether there was some facet or mind-set within the Maine settler community in the 1700s and 1800s which caused some to be almost maniacally rapacious. In 1864 Samuel

116

Boardman of Augusta related:

"Three years ago, in the month of May, in company with a friend, while passing by the lower lock of the Cumberland and Oxford Canal, in the city of Portland, our attention was drawn to the a crowd of men standing by the side of the lock, several of whom had long-handled nets, with which they were fishing, or rather dipping out fish from the water. On coming up, we saw that they were catching alewives in great numbers. It appeared that these fish, in their peregrinations along the coast, had been attracted by the fresh water of the canal, and instinctively entered it in order, as they supposed, to follow up to its source, (Sebago Lake,) but were brought to a standstill by the upper gate of the lock. The men engaged there then shut the lower gate, and commenced catching them. As soon as those of them that were confined in the lock were all caught, the men opened the lower gate again, and admitted a lot more of them, and thus a wholesale destruction of them went on."

An item titled, "A Novel Fishery," appeared in the *The American Annual Register of the Years 1827-9*, describing an event in Harpswell, Maine on October 6, 1828:

"A school or shoal of large fish, some of them between 20 and 30 feet in length, was discovered in Harpswell river, on the eastern side of Harpswell neck. A few hardy fishermen of that town discovered them, and engaged in the chase, driving them up the river and firing at them with musket balls.

"The alarm was soon communicated along shore- a whale! a whale! was the cry: --and the water was in a short time covered with boats, carrying sixty or eighty warriors to battle, armed with muskets, harpoons, broadaxes, hatchets. and whatever deadly weapon could be seized at the moment. Those who first dashed in amongst the school fired at them incessantly, and killed several, who sank in the river, where they still lie.

"The greater part were driven from the river into the cove, directly east of Harpswell meeting house, between Orrs island and Great island. The water was here shallow ; and now commenced an assault, and a method of fishery never before witnessed. The fish were known to yield a valuable oil like the whale ; the largest yielding from four to five barrels, worth thirty or forty dollars. The eagerness of attack therefore on the part of the fishermen, who were accustomed to draw up from the depth of the many fathoms the comparatively worthless codfish, may easily imagined.

"First, as became him the representative of the town of Harpswell, Mr. Curtis, a very respectable man, assaulted the largest of the school. Armed with a broadaxe, he threw himself from his boat, astride a monster 22 feet in length, and rode him a number of rods, (all the time cutting into him with the greatest industry,) before he despatched him. Mr. Dunning pursued two large fish ashore, and slipping the painter from his boat, he made a noose in it, and getting it over the head and fins of the largest, he fastened him to a tree; while snatching another rope, he slipped in over the tail of the other, and hastened to make new conquests, for it was the law of the case, that every one was to have what he could kill or catch and secure.

"The result of this adventure is, that 22 men the successful part of the company killed 71 fish, being, with those which sunk in the river the whole school. It is not known that one escaped. The blubber has been stripped off, and, it is expected, will yield 75 barrels of oil, worth perhaps from 600 to 700 dollars. Some of the Harpwell people call this fish, black fish, other pot fish. Both names are very appropriate, for the fish is black like a coal, and the head is of the form of a pot kettle.

"Dr. Mitchell, of New York, and other learned men, would say it is no fish at all, for it has no gills, and like the whale, has a heart and lungs, and warm blood, and is viviparous. It spouts water through a large spiracle or hole in the top of the head. One man thrust his fist as a stopper in the spiracle of one of the monsters, in the hope, that by confining the air, the animal would blow up, and thus be floated more readily in the shoal water; but he found himself in danger of being blown up in the air! The largest was 22 feet in length, and 18 feet in circumference; the pups still at the breast were seven or eight feet in length. "

In June, 1826 the *Kennebec Journal* reprinted a news item from a western Maine newspaper, the *Oxford Observer*:

"On Wednesday, the 31st, part of the young men of this town, and Buckfield, who had been engaged in what they denominate a *Squirrel Hunt*, met to count their game. The following is the number and kinds of animals brought in and counted:

Skunks -- 21
Rackoons -- 142
Foxes -- 10
Woodchucks -- 649
Squirrels -- 190
Bobbolinks -- 624
Crows -- 115
Hawks -- 54
Owls -- 42
Woodpeckers -- 272
Brown Thrashers -- 50
Black Birds -- 36
Cat Birds -- 32
Blue Jays -- 39

Making in the total two thousand two hundred and seventy-six. -- *Oxford Observer.*

[He who would kill a brown thrasher (or rather thrush, we believe) has no 'music in his soul.' It inhabits the thickest shades; there is more melody in its notes than in those of any other bird in our forests. We are glad to see no robins returned among the slain. The shooting of small, harmless birds cannot be sport to a benevolent and refined mind. The object of the hunt in Buckfield, however, we suppose, was to destroy animals which injure the crops of the farmer.]"

What is striking about this last news item is that all of this intentional killing and mayhem was done at a time when six days a week, 10-12 hours a day work schedules were the norm. That this was a published news item in the *Oxford Register* further suggests these killing events were a commonplace, much like a Sunday afternoon softball game. Even assuming a bit of exaggeration and puffery, the killing of 272 woodpeckers by gun in one day suggests a determined purpose with a large number of participants. What were these people thinking? Even the anonymous editor at the *Kennebec Journal* was somewhat appalled by this waste and carnage, as evidenced by his bracketed comment.

These items give a rare, unvarnished glimpse into the mind-set and everyday behavior of New England and Maine 'pioneer' settlers during the early 1800s, one which was conveniently cleansed and erased by late 19th century town historians. They help to illuminate the question of 'why' these settlers were so determinedly aggressive to wipe out the fish populations of their rivers. While alewives, shad, salmon had economic value, to be sold *en masse* at market or eaten, the massive shooting of bluejays and woodpeckers and brown thrashers in a highly organized, directed recreational killing effort on a Sunday afternoon in May had no economic purpose or value. This killing seems to have been done as recreation.

These above accounts do not mean that *all* of the settlers in the mid 1800s were so rapacious and blindingly short-sighted. Pages and pages of quill-penned documents show otherwise. A conservation ethic did exist in some people then -- or at least a knowledge of the fable of killing the golden goose.

As early as the 1770s, Silvester Gardiner's son John got into conservation, writing from Sussex, England to the Colonial Congress that his brother William was illegally damming Worromontogus Stream, destroying its enormous alewife run and depriving local people of a source of food. The people of Winthrop, Maine, at the alewife spawning ponds of the Cobbosseecontee Lakes, spent 40 years asking for enforcement against the Gardiner family to let alewives go through the small dams at Cobbosseecontee Stream in Gardiner.

The people of Phippsburg, many who were cod fishermen, complained to the Legislature there were no alewives left in the river to use as cod bait and that giant nets in Merrymeeting Bay were destroying the 'Brunswick' school of salmon headed up the Androscoggin River.

On May 22, 1829 an anonymous retired commercial salmon fisherman published a long essay in the *Kennebec Journal* in Augusta, Maine stating:[9]

"But few men now on the stage appear to have a knowledge of the superabundance of these fish as far back as the period before the Revolution. If my memory does not misgive me, a Mr. Rogers and his company, seven all in the year 1784 or 85, at Hunnewell's Point, exclusive of Fox Island, took in set nets between eight and nine thousand salmon. The average weight of each was 20 pounds of the first shoal, and the last 18 pounds. When I owned the same fisheries, the salmon were two to three pounds lighter. Where Rogers caught a thousand, my fishermen had need be industrious to get 100."

This 1829 essay is striking because the author is already calling for an overhaul of state laws to *restore* the Kennebec's alewife, shad and salmon fisheries: "If therefore the Salmon, Shad and Alewife fishery, even in that part of the Kennebec could be restored, it would give a

---

9   This item carries an editor's note stating that it was originally published in the *Wiscasset Citizen*. At this time, Maine newspapers did very little original reporting, but instead mostly reprinted squibs of material from other newspapers. The *Kennebec Journal* is almost unique in that all of its editions from 1825 onward are preserved on microfilm at the Maine State Library. In contrast, most of the Maine newspapers cited in it were of such short duration that no copies survive today.

source of wealth to the State about equal to the amount of the State tax."

These are the reasons why the chain pickerel came to be spread across Maine.

The Alewife who went to the U.S. Supreme Court

On a May day in 1739, a year about equidistant from the Pilgrims' landing in Plymouth, Mass. and Maine becoming a state, a man from southern Maine boarded a sloop from a dock in Saco Bay to a dock in Boston. The purpose of his trip was to lodge a complaint against a retired army colonel for building an impassable dam on the Presumpscot River in Falmouth, Maine downriver from where he lived.

Upon reaching Boston, the man walked the short distance from the harbor docks to the State House. He sought and received an audience with the Governor of the Massachusetts Bay Colony, Jonathan Belcher, who was King George's appointed steward of the Crown's Colony in New England. There he aired his grievances.

A short, surviving summary transcript of the meeting in the Massachusetts Archives shows the man said:

"Your Excellency, when the treaty was sealed was pleased to say that if any thing should happen that we could not understand or did not approve of we should inform your Excellency of it: and what we are most aggrieved at is that the River Pesumpscut is dammed up so that the passage of fish which is our food is obstructed, and that Col. Westbrook did promise about two years ago that he would leave open a place in the dam and that the fish should have a free passage up said

123

river into the Pond in the proper season but he has not performed and we are thereby deprived of food."

Soon after this meeting, a letter dated August 13, 1739 and signed by John Willard, Governor Belcher's executive secretary, was sent to the dam owner, Col. Thomas Westbrook. The letter reads:

"Sir,

I am directed by His Excellency & the Council to acquaint you that divers Indians inhabiting on Pesumpscot River have complained that by the dams built on that river the course of the fish is stoped & they are thereby deprived of a great part of their subsistance, that upon your first building the dam a passage was made therein for the fish & kept open in the proper season, but of late that passage has been wholly stoped up. I am further to acquaint you that the Governor and Council apprehend it but reasonable to leave open a sufficient passage for the fish and this they may expect may be done that no further complaint may be made in this matter and rather because the deed of President Danforth to the Town of Falmouth does make an express saving of the rivers. It is also desired that you would take care and give orders that the people of Pesumpscot River treat the Indians kindly that come hither.

J. Willard."

The man's name is recorded only as Polin, an Indian. Secondary documents show that Polin was a chief or Sachem of a group of Native Americans who lived along the Presumpscot River and Sebago Lake. His tribe is variously ascribed as the Arasagunticook, ie. the Native Americans who lived along the coast of southern Maine and up its river valleys.

Polin brought a legitimate and almost airtight claim to Gov. Belcher in 1739. By this time, the land deeds made between Indians and English were scrupulously documented and filed in the colonial deed archives. English settlers had deeds of record for land sales by Indian

Sachems along the Presumpscot River from Casco Bay upstream as far as Ammoncongin and Saccarappa Falls, at what is now Westbrook, Maine. But the deeds went no further up the river toward Sebago Lake. The rest was legally recognized Indian Land.

Gov. Belcher's assistant notes that "the deed of President Danforth to the Town of Falmouth (now Portland, Maine) does make an express saving of the rivers." This meant, in the legal structure of the times, that the King of England, and therefore the Massachusetts government, retained ownership of navigable waterways like the Presumpscot River, and that Col. Westbrook had no legal right to dam the river except under conditions approved by the colonial government, which in this case, included a condition that Westbrook leave a place in the dam for fish to pass. The dam was located at Presumpscot Falls, at the river's head of tide.

This exchange marked the first of a rolling series of disputes over fish passage on the Presumpscot that continued for the next 250 years and in 2006 ended up before the nine Justices of the U.S. Supreme Court.

In 2006 the Supreme Court took up the question, raised by the South African corporation which owned five dams on the Presumpscot River, of whether the U.S. Clean Water Act applies to hydroelectric dams on U.S. rivers.

Unlike any other U.S. or state court, the U.S. Supreme Court gets to pick and choose which cases it wishes to hear and which it does not. If someone wishes the Supreme Court to hear their case, they must file a writ of certorari to the Court. The Court declines to hear nearly all of the thousands of writs it receives each year and at most selects a few dozen each year which the Court believes raise core questiosn about the meaning of the U.S. Constitution and federal laws.

The question of whether the U.S. Congress intended the Clean Water Act to apply to hydroelectric dams and their effect on fish, esp.

sea-run fish like alewives, had been bandied about by federal courts and the Supreme Court for many decades prior to 2006. The U.S. Supreme Court could have denied the writ of the SAPPI corporation, the owner of the Presumpscot dams, on grounds the question had already been settled. But instead the Court decided to take the question head-on, presumably to settle it once and for all.

The reason the SAPPI corporation decided to go to the U.S. Supreme Court was that it did not want to have to spend its own money to install fishways on its dams on the Presumpscot River for alewives, blueback herring, American shad, Atlantic salmon and American eel. To achieve this goal of avoidance, SAPPI decided to argue to the U.S. Supreme Court that the State of Maine had no legal authority to tell them to put in fishways at their Presumpscot River dams. In effect, SAPPI argued that it alone owned the Presumpscot River and the public had to accept the river in whatever condition SAPPI had left it, which in this case meant without any of its native sea-run fish.

For myself and others involved in the case, SAPPI's position was as much comical as it was chilling. Here was a multi-billion dollar corporation based in South Africa attempting to overturn several thousand years of settled western law on a fairly small river in one of the least populated states in the United States just to avoid spending a mind bogglingly minuscule amount of their annual gross revenues to put in a fishway at their dams.

Wait. Several thousand years of settled law? The question of who owns things like lakes, rivers, oceans, seas and beaches has been in dispute for millennia. The Roman Empire, as it did with so many things, were among the first people to turn this question into the novel concept of written law.

The Romans devised the concepts of the *jus publicum* and the *jus privatum*. This division recognized that things like rivers, lakes and coastlines, like water and air, are resources that everyone depends on and are used by many people for many purposes, some harmonious with each other, and others not so harmonious. Inevitably tensions will arise. So the Romans stated as a first principle that the government, ie. the sovereign, is the ultimate owner. But the sovereign, as the owner, could divvy out to private citizens a permission or license to use lakes, rivers and coasts for certain specified purposes, such as putting a dock in a cove to load and unload boats.

Under the Roman concept, the conveyance of a license or permission for a person to use a sliver of coastline for a dock was an exercise of the *jus privatum*, meaning a 'private use" of a public thing. This meant, in practical effect, that if someone got permission to build a dock along the seashore, it was their private dock, but the sand and mud its pilings were sunk into remained the sovereign's, ie. part of the commons, and if someone wanted to walk under the dock along the beach, they could without interference. This was called *jus publicum*, or the 'public use,' or more exactly, that portion of the property right the sovereign chose to retain for its own and the public and not relinquish to a private party.

By the 1700s, the Roman concept of the *jus publicum* and *jus privatum* had survived centuries of English case law with little change. This body of settled law was imported to New England in the 1600s and 1700s and grafted onto U.S. law when the United States broke free from England. This was the body of law cited by Mass. Gov. Belcher to Thomas Westbrook in 1739, where he states that the State, the Colony, the Crown, the King, or whatever you wish to call it, made an "express saving of the river," meaning Gov. Belcher had the legal authority to order Col. Westbrook to let fish go past his little dam. Relying on 2,000 years of settled western law, Belcher said to Westbrook, "you don't own the river. We do."

Two hundred fifty years later, in 2006, the SAPPI corporation sent a slew of attorneys to Washington, D.C. to inform the U.S. Supreme Court that its *jus privatum* of the Presumpscot River so greatly outweighed the *jus publicum* that the State of Maine couldn't tell them to build fishways at their Presumpscot River dams.[10] The Justices were not convinced. They ruled 9-0 against SAPPI. A native of nearby New Hampshire, Justice David Souter, wrote the Court's unanimous opinion in what is now called *S.D. Warren v. Maine Board of Environmental Protection*. So what does it say?

Well, in a way, Souter started with Article One, Section 8 of the U.S. Constitution, which states Congress has the authority to regulate interstate and foreign commerce. This is called the "Commerce Clause." Next we go to an 1824 ruling by the U.S. Supreme Court in which the justices decided, based upon the Commerce Clause, that the U.S. Congress, rather than the state of New York, had ultimate authority in regulating ships and ship traffic on the Hudson River. This initial decision led later federal courts to define all but the smallest of the nation's rivers as 'navigable waters' -- even if they can only be travelled by canoe.

In the 1970s, a majority of members of the U.S. Congress decided that many of the nation's waterways were so filthy with pollution that something needed to be done to clean them up. So they passed what is now commonly called the U.S. Clean Water Act (CWA). The authority of Congress to pass the CWA comes directly from the Commerce Clause of the U.S. Constitution.

Now the CWA is pretty long, well ... really long, and requires all kinds of stuff. One thing it requires is that each state enact legal water quality standards for its lakes and ponds. These standards can vary widely by state and by waterbody but they all have to meet certain minimum standards set forth in the CWA. A second thing the CWA does is to require anyone who needs a federal permit for a project in a

---

10 Purists will note SAPPI didn't exactly say this, but I'm not a purist. The sole intent of SAPPI's challenge was to prevent Maine from protecting important public uses of the Presumpscot River, like ensuring its native fish could live in it. Had the State of Maine not required fish passage, minimum flows and dissolved oxygen levels necessary for fish life in its water quality certification, SAPPI would never have challenged the certification to the U.S. Supreme Court in the first place.

particular state to get a piece of paper from the state which says the project will not violate the water quality standards of the state. This is called 'water quality certification.' And if you don't get that piece of paper, the federal agency can't give you a permit.

Another fun law that Congress passed, this time in the 1920s, is called the Federal Power Act (FPA). It is also based on the Commerce Clause and says that anyone who wants to own or build a hydroelectric dam on the nation's rivers must get a permit under the Federal Power Act. In the 1920s, the whole concept of electricity was just a few decades old as was the idea of damming up rivers to generate electricity. It was the new fad. So lots of dam owners who used to use the dams to turn waterwheels and gears attached electrical generators to the gears and voila, free electricity. In passing the FPA in the 1920s Congress gave one free license to anyone who, at the time, had a dam rigged up to generate electricity. These licenses had terms of 20 to 40 years, the idea being that Congress wanted to encourage the construction of hydroelectric dams, it being the new fad and all, and long license terms made it easier for dam owners to obtain bank financing to build or improve the dams because the cost of construction could be spread out over several decades. So what's all this got to do with alewives?

In the late 1990s, the federal licenses the SAPPI corporation had long possessed to run five of their dams on the Presumpscot River were about to expire, so they had to apply for new ones. But under the Clean Water Act, they also had to get a piece of paper from the State of Maine saying the dams would not violate any state water quality standards for the Presumpscot River.

But instead of just applying for this piece of paper from the state, SAPPI's attorneys also sent the state a nasty-gram which said they didn't believe they needed a piece of paper from the state because everyone knows that dams don't cause water pollution because they don't put anything in the water besides the same water that's above the dam. In response, the state said, sorry you feel that way SAPPI, but if you don't apply for your water quality certification, you can't get your federal license and then your dams will have to be all ripped down. So SAPPI begrudgingly sent in their application but also said they were going to sue the state all the way to the U.S. Supreme Court.

But that was not all that was up SAPPI's sleeve. Because they had a strong sense the state was going to order them to build fishways at all of their dams they asserted that fish like alewives never went past their dams even before they were built, because the dams were all built on natural falls too steep for fish to get over. And if the fish never went past the dams naturally, they argued, the state had no authority to tell them to build fishways.

Now this was a head scratcher for the state, since dam owners never made claims this bold. But SAPPI's attorneys were smart and had done their homework. At some point they had a minion go to a library and look up every single book they could find about the history of the Presumpscot River and discovered there aren't that many. More importantly, all the books they did find were all written 150 years or more after the river had first been dammed by Col. Thomas Westbrook at its mouth in the 1730s. Not a single book, letter or diary existed which described the river before it had been impassably dammed. So SAPPI said to the state, "We assert sea-run fish never went past the sites of our dams and you don't have a shred of proof they did, so you lose and we win."

At this time, around 1999, a couple conservation groups had entered the fray and hired a researcher to look for any and all historic documents which might refute SAPPI's claim. After about a month of poking around libraries, she glumly reported back that aside from a few vague references to fish in some flowery local history books from the late 1800s, SAPPI was basically right: not a single reliable account existed describing fish like alewives or salmon going up and over the falls where SAPPI's dams were built. This was not surprisingly, she noted, because nobody except a few Indians lived on the river when it was first dammed in the 1730s, and they were all driven out and/or shot soon after the first white settlers moved in. For the next two years, as the dam licensing crawled along, it started to look more and more like SAPPI held a legally unbeatable hand, and all for the lack of a single piece of paper from someone 250 years ago who saw sea-run fish up past SAPPI's first dam and took the time to write it down.

During this period of gnawing glumness and impending doom, I happened to be at the Maine State Library in Augusta, quasi-randomly going through microfilm spools looking for information about the Kennebec River. One of the spools was labelled "Massachusetts Archives" and I put it in the machine. About halfway through the spool, a scratchy and barely legible document popped up in the viewer. It was originally written with quill pen and the lines slanted all over the page. It was a petition to the Massachusetts Legislature from a dozen or so people who lived along the Presumpscot River in August 1776 asking the Legislature to require the dam owners to let alewives and salmon and shad to go past their dams. It read:

"The Petition of the Towns of Cape Elizabeth, Windham, Gorham and Pearsontown in the County of Cumberland

Humbly Shew

That the said Towns lay bordering on Presumscutt River, so called, and for many years after the Settlement of this Eastern Country were plentifully supplied with Salmon, Alewives, Shad & other sorts of Fish that frequented the said River in great abundance it being peculiarly commodious for the Spawn & increase of Fish by reason of a large pond called Sebago or Sebacook which extends upwards of thirty miles from the mouth of the said River as far as Pondicherry as also the many branches of said river that used to bring a plenty of aforesaid Fish near to many of our doors, your Petitioners further shew that by reason of several Mill Dams being built quite across said River, without leaving a sluice way for Fish to pass up, as by Law is directed, and since the said Mill Dams have been erected on the said River the passage of all kinds of Fish as aforesaid has been totally obstructed & stopt in their course up said River to the great prejudice of many back Towns which

131

depended (in their Inland state) on the said River for a part of their support, as also to the prejudice of all the Inhabitants of the Sea Coast near the mouth of said River by causing a scarcity of Codfish, Haddock, and many kinds of Fish that frequent the mouths of such extensive Rivers after a quantity of small bait that abound in such places. And our fishing on the Banks as well as on our Coast off shore being in a great measure impracticable by reason of the Enemy's cruisers that infest our Coast, reduces us to the necessity of adopting some method whereby the fish may come to us. And the Laws of this Colony have been found ineffectual hitherto for the removal of your Petitioners cause of Complaint, Wherefore your Petitioners pray Your Honours to take the matter of our Complaint into your consideration and Grant to your petitioners such relief as in Your great Wisdom & Clemency You may Judge meet & your Petitioners as in Duty bound shall ever pray. August 22, 1776."

A month or so later, I was again wasting time at the library, idly thumbing through a UMaine history student's Masters Thesis about the Penobscot Indians in the early 1800s. The footnotes kept referring to documents at the Maine Archives with a catalog code that read "Legislative GY" followed by a number. So I went upstairs to the Maine Archives and asked them what that code meant. The archivist on duty that day, Anthony Douin, a 60ish year-old local historian I knew from town, said, "Oh yeah. That means Legislative Grave Yard. Those are petitions for laws submitted to the Legislature that never got enacted into law. We have a whole index for them over here on the shelf. Just go through the catalog, pick out the ones you want to look at and we can bring them up from the storeroom."

Jackpot !!! Little did I, or anyone else involved in the re-licensing know, that the Maine Archives scrupulous maintained and guarded the equivalent of a Dead Letter Office going back 250 years which contained almost every letter someone in Maine had sent to the Legislature complaining about something and asking them to do something about it. And the letters were all original. You could hold them, feel the paper, examine the faded ink and water stains, admire the penmanship or lack thereof.

As a week or so in the Maine Archives revealed, there had been quite a long and loud stink in the late 1700s about the Presumpscot River being dammed up and the fish not getting upstream, and it was (thankfully) very well documented. Going back to the microfilm spools at the State Library I found one small reel for a newspaper called the Falmouth Gazette and Advertiser, published only for a few years in the 1780s. Jackpot again !!! The front page of several editions had LEGAL ADS notifying Presumpscot dam owners to 'show cause why they should not build sluiceways at their dams' for sea-run fish. One edition devoted its whole right side column to the full text of a new law, passed in 1785, specifically requiring fish passage for alewives, shad and salmon on the Presumpscot River. And it was signed by none other than Samuel Adams himself, who was then President of the Massachusetts Senate. The alewives of the Presumpscot River had the Founder of the American Revolution itself on their side. As I hit the 'print' button on the microfilm machine, I thought, "Take that, you giant South African corporation !!!" and giggled maniacally to myself.

So once all of these dozens of 1700s documents were properly collated, catalogued, copied and affixed with a very official and not excessively gloating Executive Summary written by the legal counsel for American Rivers and Friends of the Presumpscot River, they were shipped off to the state, the feds and the D.C. law firm who represented SAPPI. And that was the last we ever heard from SAPPI about alewives and salmon not historically going above their dams on the river.

Now we were back to SAPPI saying the state couldn't make them build fishways because the Clean Water Act has nothing to do with dams. They also said crazy stuff like as long as the water above their dams was clean enough for alewives and salmon to live in, it was irrelevant if they could actually get above the dams and that fish only need dissolved oxygen on an 'average daily basis' and not when they take their next breath.

This legal freight train rolled into Maine Superior Court, who ruled against SAPPI, then to the Maine Supreme Court, who ruled against SAPPI, then to the U.S. Court of Appeals, who ruled against SAPPI.

SAPPI's attorneys, down by several runs in late innings, decided to swing for the fences rather than trying for a bunt single to advance the runner. So as promised, they petitioned the U.S. Supreme Court to hear why they should not let alewives go above their dams. The U.S. Supreme Court decided to hear the case. Somewhere today there's a forest that gave up quite a few of its trees to produce the mountain of legal paper which soon spewed forth to answer the question, as re-phrased by the Supreme Court: "Do hydroelectric dams result in 'discharge' under the meaning of Section 401 of the federal Clean Water Act?"

SAPPI's linchpin was a 2004 case by the Supreme Court involving the Miccosukee Indian Tribe and the pollution of the Florida Everglades. There, the tribe sued the State of Florida, claiming that the pumping of polluted water from an artificial canal to an artificial reservoir in the Everglades was a discharge of pollutants under the Clean Water Act that required a wastewater discharge license. SAPPI hammered on a quotation from the decision, wherein the Supreme Court ruled against the Miccosukees and agreed with a lower court judge who reasoned, "If one takes a ladle of soup from a pot, lifts it above the pot, and pours it back into the pot, one has not 'added' soup or anything else to the pot." Hydroelectric dams, SAPPI argued, were the perfect analog to taking a ladle of soup from a pot and pouring it right back into the pot. Because its dams do not add pollutants to the water, SAPPI argued, there is no way their dams could be called a 'discharge' of pollutants and fall under the Clean Water Act.

But Justice David Souter noted the CWA is quite long and has many sections and purposes. Some sections strictly deal with adding 'pollutants' to a river, but others, such as the section involving federal permits for things like dams, deal with the broader concept of water 'pollution.' While adding 'pollutants' can certainly cause water 'pollution,' Justice Souter noted the Court had recently ruled that a riverbed dried up below a dam is also a form of 'pollution' since fish cannot live in a dried up riverbed. So he concluded that, "When it applies to water, 'discharge' commonly means a 'flowing or issuing out.' ... In fact, this understanding of the word was accepted by all Members of the Court sitting in our only other case focused on Section 401 of the Clean Water Act."

Strike three swinging. Ball game over. As Harvard University Law student Madeline Fleisher wrote in the 2006 edition of the *Harvard Environmental Law Review*:

"The *S. D. Warren* decision will have important ramifications. On the most basic level, *S.D. Warren* will affect the future of more than a thousand non-federal hydropower dams across forty-five states, many of which will have their licenses come up for renewal in a new era of more aggressive state environmental regulation. A narrower interpretation of the term "discharge" by the Supreme Court could have left the states with no way to regulate the impact of some dams on water quality ... "

In a way that many judges do not, Justice David Souter saw the forest for the trees and protected the *jus publicum*, limited its engulfment by the *jus privatum*, and resisted giving credence to "the exception that swallows the rule." In doing so, he affirmed the decision made by Massachusetts Governor Jonathan Belcher one day in May in 1739 after meeting with Chief Polin of the Presumpscot, who came down on on a sloop from Maine to talk about alewives. About ten years after this meeting in Boston, Chief Polin was shot and killed with a musketball by English settlers along the banks of the Presumpscot River, which by then had five impassable dams on it.

Alewives cause
heartburn, strife and idleness

One thing that stands out in the hundreds of pages of legislative fisheries petitions submitted by citizens to the Maine Legislature during the 18th and 19th centuries is that virtually all of them are from citizens asking the Legislature to ensure that alewives, shad and salmon could get upstream past mill dams.

The one exception I found is from Warren, Maine on the St. George River, one of the only rivers in New England which has never lost its native, aboriginal alewife run. The dam owners' petition from the winter of 1837 reads:

"To the Honorable Senate and House of Representatives in Legislature:

Your petitioners, Inhabitants of the town of Warren, respectfully represent that the law regulating the taking of fish in the St. Georges River in said town requires that the dams across said river shall be opened during the season when the fish pass up and down said river. This is a great injury to the owners of the mills situated on said river and to all having business at said mills."

That the law, requiring the dams to be opened as aforesaid, prevents the vacant water privileges on said stream, which are among the first in the state, being occupied for factories and various other machinery, depriving the inhabitants of those advantages which Nature has provided them, and thus retarding the growth and prosperity of said town and the good citizens of the surrounding country. "

That the right of taking fish in said town or the law regulating the same has become a bone of contention among the people and a prolific source of litigation, some contending that they have a right to take fish in the navigable waters in said town, and taking them accordingly, notwithstanding the law, others owning farms bordering on the river, contending that they have a right to fish in the waters upon their own land, and fishing accordingly, the law to the contrary notwithstanding. Thus are generated heartburnings, strife and lawsuits. The fish for years back have much diminished, and they do not when obtained half pay for the trouble and expense of taking and curing them to say nothing of the great waste of time by our citizens in congregating and waiting about the fishways. On the whole your petitioners are satisfied that it would be for the benefit of the citizens of this and the neighboring towns to have the law aforesaid repealed, and they do hereby respectfully request the Legislature do the same."

Apparently, the people of Warren quickly got wind of this petition, called a Town Meeting to discuss it, and fired off their own letter to the Legislature, dated February 6, 1837, which reads:

"At a legal meeting of the Inhabitants of the town of Warren qualified to vote in town affairs, Voted that our Representative to the Legislature be instructed to oppose any petitions that may be presented to repeal the law regulating the Shad & Alewife Fishery in the town of Warren. Voted that our Representative be furnished with a copy of the above vote by the Clerk.

Attest, Stephen C. Burgess, Town Clerk to Edward O'Brien, Esq."

Since these two documents are in the 'Legislative Grave Yard' of the Maine Archives (Box 110, Envelope 9), the Legislature apparently refused to act on the dam owners' petition and observed the town's vote at Town Meeting to leave the existing requirement for fishways in place on the St. Georges River, and in doing so, preserved the alewife run, which today totals over 100,000 fish.

The value of these two documents today is that they provide a very rare eyewitness glimpse into the contentious dynamic that existed on virtually every river in New England where the interests of sea-run fish and the people who relied on them collided with mill dam owners who resented having to build fishways.

Here, the dam owners of Warren throw in everything but the kitchen sink. First, they argue that the cost of maintaining and repairing the existing fishways at their dams is a big, expensive headache for which they receive no benefit. Next they claim that the legal requirement for fishways at any new dams on the St. Georges River discourages people from building any new dams on the river. Then they claim that all of the various parties involved in the alewife fishery argue and fight so much about where people can fish and how many they can take that it occupies far too much of everyone's time. Then they claim that the time and cost in catching and curing the fish (ie. smoking them) outweighs the monetary value of the fish. Then they claim that in recent years the alewife run has shrunk to the point where it's not really worth preserving. Finally, they pull out the old Calvinist card and state that during the alewife run far too many people in town spend their time gawking at the fish and catching them rather than doing productive, useful work, like presumably, working at mill dams.

What is downplayed in the dam owners' petition to the Legislature is that they are advocating for the entire eradication of the native fish runs of the St. Georges River on purely economic grounds, meaning of course, their personal economic grounds, and the complete

138

destruction of a commercial and sustenance fishery that had been maintained for a century since European settlement and for millennia by Native Americans. This is the 'time marches on' and 'it's the price of progress' and 'fish are quaint, but not necessary' argument. Little heed is given to the economic interests of the people in town who actually catch and sell and eat alewives. Apparently some peoples' economic interests are more important than others.

The second fascinating facet of this 1837 petition is that dam owners in New England today still make exactly the same arguments, excepting maybe the 'idleness' part. In the late 1980s for example, attorneys for the S.D. Warren paper company of Westbrook, Maine argued before the Federal Energy Regulatory Commission that putting water back into the 1.25 mile section of the Presumpscot River de-watered for a century by their Eel Weir Dam was a bad idea because if water was restored to the river it would cause a dangerous safety hazard for anglers, who might fall in and drown while trying to catch a fish.

They Dammed Paradise
and Put Up a Fishing Spot

In the long history of alewives in Maine, where innumerable laws to preserve them have never been enforced, it is their hard luck that the only law purposefully passed to make them go extinct is the only one enforced to the letter.

Like all of New England's large rivers, the Schoodic (now called St. Croix) in easternmost Maine, had its alewife, shad and salmon runs reduced to near extinction by the 1830s by impassable dams. Atkins & Foster stated in 1867:

"The St. Croix was formerly very productive of salmon, shad and alewives. Perley, in his report on the fisheries of New Brunswick, states that the average catch of salmon at Salmon Falls, Calais, was 18,000 annually. Gaspereaux (alewives) came in such quantities that it was supposed they could never be destroyed. The number of shad were almost incredible. The fisheries did not diminish until 1825. Until that time the dams had fishways; but in that year the Union dam was built without a fishway, and the fisheries instantly fell off."

These events led to the transfer of chain pickerel, then white perch and then the smallmouth bass to the vast lake system along the U.S.-Canadian border to replace what had been lost. In 1868, Atkins and Foster stated of chain pickerel:

"From the Penobscot they were transferred to the St. Croix twelve or fourteen years ago, by carrying a number of them forth from Baskahegan lake to Little Musquash lake. From the latter lake they have spread into all the St. Croix waters; the trout (*Salmo fontinalis*) has given place to them on the lower lakes, but the Schoodic salmon and togue stand their ground better. From the St. Croix they have been transferred to Meddybemps lake, on the Denny's river, and have in some way obtained a footing in the East Machias River, where they have become abundant."

Smallmouth bass were the third fish, after pickerel and white perch, to be spread widely by a cadre of Maine 'bucket' naturalists, many of whom, such as Ezekiel Holmes, were at the time the state's leading and most learned naturalists. Unlike white perch and pickerel, the smallmouth bass is foreign to the Atlantic seaboard and is native to the Mississippi River drainage. Boardman (1864) states:

"It is a very common species in all of the Canadian lakes, except Superior, and on the St. Lawrence river, Lake Champlain and its tributaries. It is also found in several localities in the interior of New York, and has been introduced into some of the waters of Connecticut, Massachusetts and New Hampshire. From Mr. S.T. Tisdale of East Wareham, Massachusetts, we have the following: 'They were introduced to the waters of this region by myself, in 1850, '51 and '52, to the extent of some two hundred, with which I stocked some in ponds in this vicinity. They were procured at Saratoga L ake, N.Y., and brought here.'"

The construction of the Erie Canal allowed smallmouth to spread from Lake Erie down the Mohawk River in the Hudson River drainage. Intentional distribution of smallmouth east of the Hudson

141

began in the 1830s through a network of self-styled naturalists and learned, gentlemen who considered fish transplantation a scientific hobby. By the 1850s the smallmouth were the new fad among New England bucket naturalists, as Atkins & Foster (1868) state in a chapter titled, *Dissemination of a Valuable Fish Species*:

"Black Bass. -- This superior fish was described on page eighty-six of the report for 1867. A further acquaintance justifies all that was said in his favor, and relieves him in part from the accusation of being a gross feeder upon his fellow fishes. He is by no means as voracious as the pickerel, -- probably not more so than white and yellow perch. All things considered, it was believed that it was very desirable to introduce this species to our waters ...

"Thirty four black bass were brought from Newburg, NY and deposited in Duck Pond in Falmouth. A large lot was delivered in Bangor, by E.S. Woodford, of West Winsted, Connecticut. The greater portion of these, about 60 in number, were put into Fitz pond, in the town of Dedham; 28 into Newport pond, and ten were sent to Phillips (Pond). Previously, in the month of August, Cochnewagan Pond, in Monmouth, and Cobbosseecontee Lake, in Winthrop and adjoining towns were stocked with bass at the expense of the Oquossoc Angling Association. The fish for this purpose were furnished by Walter Brown, from his private ponds, in Newburg (NY), and were brought to their destination by Geo. Shepard Page, president of the OAA. Mr. Brown also furnished, gratis, the bass with which Duck Pond was stocked."

The invention of the railroad during this time coincided with the equally novel concept of the 'sport angler,' a fairly well-to-do person who found recreation by catching fish on a rod and line for sport as much as for food, often in exotic locations serviced by an equally new concept, the fishing lodge, hotel and resort. These facilities were actively promoted by railroads to increase ridership. Angling clubs and asssociations, another new invention, became the funding source necessary to attempt shipping live, adult smallmouth bass from ponds in the Hudson Valley for release in ponds in New England. This can be seen by Atkins & Foster's statement that some of the earliest transplants of smallmouth bass to Maine were organized, directed and funded by the Oquossoc Angling Association, a well-heeled gentlemens club of sport anglers who fished for the monstrous native brook trout of the Rangeley Lakes on summer excursions. By the 1870s it appears certain that smallmouth had been introduced to the St. Croix River chain of lakes, where they thrived and created a new and highly desirable sports fishery to visiting, paying angler-tourists.[11]

For next 100 years the St. Croix River lakes became one of the world's most well-known and popular destinations for sports anglers to catch smallmouth bass in a wilderness, but comfortably appointed setting, with the benefit of native guides, well-cooked meals and a roaring lodge fireplace. The river's native sea-run fish, the Atlantic salmon, the shad, the alewives, were long forgotten, as the old timber crib dams which blocked their ascent remained firm.

In the 1960s a brief spurt of interest in restoring sea-run fish caused the U.S. government to fund new fishways on the dams of the St. Croix, but a particularly old one at the head of tide on the Canadian side of the river, at Milltown, was so dilapidated that few fish could use it. Finally, in 1981, after much of the raw paper mill waste which fouled the lower river had been abated, the New Brunswick Power Company rebuilt the Milltown dam fishway, and the alewives started coming back. With access to most of the watershed, alewife numbers skyrocketed from a few tens of thousands in the early 1980s to 2-3 million by the late 1980s. With just the fixing of this one old fishway, the St. Croix by the late 1980s had the largest native alewife population on Earth -- but at 2-3 million fish it was stilll shadow a fraction of its original size of 21 million fish.

At this time, a multi-national paper corporation, Georgia-Pacific, controlled most of the large dams on the Saint Croix, including

---

11 Henry Thoreau was one of this new generation of middle-class, southern New England eco-angling tourists, from which he crafted his book, "The Maine Woods,' which began as a popular magazine writing assignment.

a storage dam at Spednic Lake, a sprawling natural lake straddling the U.S. -Canadian border. Each winter during the 1980s, G-P would draw down the level of Spednic by up to 15 feet by opening the gates at the dam in the winter. This water flowed downriver and turned the turbines of G-P's hydro-dam in Woodland, Maine. This increased the amount of power the Woodland Dam could generate, especially in the winter, when river flows are at their lowest (everything's frozen) and electricity prices are high. These draw-downs caused about 3,000 acres of Spednic Lake, the shallow parts, to dry up and freeze up solid every winter. Everything in the shallows of the lake was killed. Nobody said anything.

During this same period, people who fished for smallmouth bass in Spednic Lake, including part-time guides employed by 'sports' from away to fish for smallmouth bass, noticed a falling off in the number of smallmouth bass in the lake in the summer compared to earlier years. They wondered if the increasing numbers of alewives in Spednic Lake in recent years had something to do with what appeared to them to be a decline in smallmouth bass numbers. They brought their concerns to the local state fisheries biologists, including one named Michael Smith. In the mid 1980s, Smith donned scuba gear and took a number of swims in Spednic Lake. In an account of his dive given to Roberta Scruggs of the *Maine Sunday Telegram* in 2001, Smith said, "I'd dive in the water and I'd see a school of alewives that would be 50 to 100 feet wide and 300 to 400 feet long. Nothing but alewives. Everything else was gone."

In 1995 the smallmouth bass anglers and guides got their local legislators to sponsor a bill to eradicate all of the native alewives in the Saint Croix River by blocking their passage at the river's fishways. The bill became law and had its intended effect. The Saint Croix River alewife run fell from about nearly 3 million adults in 1987 to only 900 adults in 2002. What happened?

What has happened on the St. Croix watershed since the 1980s was tragic and completely avoidable. The massive seasonal water fluctuations at Spednic Lake caused by the Georgia-Pacific corporation's dam manipulations created a hostile and unpredictable environment for everything living in the shallows of the lake. G-P got away with it for decades because the bass population was just resilient enough to take it, but was teetering on the edge of disaster. The St. Croix is close to the northernmost limit of the smallmouth bass, with short, cool summers and long, cold winters. Research indicates that baby smallmouth bass, born in June, must grow to a critical length by fall or they do not survive the winter. This means that the availability of food items in the months of July and August and early September are critical to that year's new generation of bass. The massive winter lake level drawdowns at Spednic, which de-watered the shallows down to a depth of 10-15 feet, had a devastating effect. Natural lakes do not fluctuate by 15 feet in a year and natural lake organisms have no evolutionary adaptations for such dramatic seasonal changes.

In the 1980s and 1990s the consensus of the more thoughtful fisheries biologists in Maine was that the large and quite sudden influx of native alewives to Spednic in the mid 1980s caused a biological 'tipping point' to be surpassed. In their already debilitated condition, the lake's shallows could no longer support sufficient invertebrate life, primarily the zooplankton *Daphnia*, to allow for adequate mid-summer growth of baby smallmouth bass, baby alewives and all of the other fish species which consume *Daphnia*.

In 1999, Fred Kircheis, a senior research fisheries biologist with Maine DIFW, now retired, gave me this assessment of the situation at Spednic Lake during the 1980s, "With the annual winter drawdowns, you have a large loss of organisms in exposed parts of the lake, everything from freshwater mussels to aquatic insects and aquatic plants. This greatly reduces the productivity of the lake and its ability to support fish life."

A plan was devised in 1987 by Maine DMR, Kircheis and other biologists which called for keeping alewives out of Spednic for a few years as a stop-gap measure, making Spednic Lake catch and release only for smallmouth, while at the same time, working on getting

Georgia-Pacific to reduce the severity and magnitude of its winter drawdowns. By the early 1990s, G-P had reduced its winter drawdowns to no more than 10 feet (still well more than natural), the catch and release rule remained in effect, and alewives were not let into the lake. By the mid 1990s all data showed the lake's smallmouth population had stabilized and was on the rebound.

These improvements lent some credence to the hypothesis that recent large influx of alewives to Spednic, in concert with the drawdowns, had created a 'tipping point' in the lake's ecology which was observable in a reduction in juvenile smallmouth bass recruitment. However, because the post-1987 changes tried everything at once (going to catch and release for smallmouth bass, reduced winter drawdowns, no alewives in the lake), it was impossible to say if one, some, or all of these changes were the driver of the stabilization of the bass population observed in the early 1990s.

Had the 1987 plan had been allowed to continue as planned, by the mid 1990s Maine DMR would have begun allowing alewives back into Spednic Lake in controlled numbers while Maine IFW monitored for any signs of a crash in juvenile bass recruitment and the story would end right here. But that's not what happened.

What happened instead was the smallmouth bass fishermen and guides decided they wanted alewives out of the St. Croix River completely and forever. In 1995, the Maine Legislature, acting on the guides' request, passed emergency legislation ordering the state's fisheries commissioners to ensure that no alewives could enter any part of the St. Croix River. What began in the mid 1980s as a very localized and unique problem on Spednic Lake caused by artificial 15 foot drawdowns had by the early 1990s turned into a *jihad* against a native fish species which has lived in the St. Croix since the last Ice Age.

It's All My Fault ... Really.

In the early 1990s, my dad and Joe Cardoza and Ralph Archibald and Tony Boynton drove six hours from Easton, Massachusetts each October to Grand Lake Stream to Tony Boynton's cottage to fish for the fabled Schoodic salmon. I accompanied them for several years, driving to Old Town from Augusta, and then along the dusty Stud Mill Road for hours to spend some quality time with my dad and to kick his ass in some *mano a mano* flyfishing over very fussy native salmon.

At the time I was also writing a monthly column, "The Maine Conservationist," for the *Maine Sportsman* magazine. Even though I was an accredited outdoors journalist in Maine, and fishing at Grand Lake Stream with my dad and Joe Cardoza each fall, I had no idea that an alewife war was in full swell on the St. Croix, which had its focus right at Grand Lake Stream. Neither did my dad, Ralph, Tony or Joe. We were too busy fishing, frying linguica in a skillet and telling extremely off-color jokes. Such is the life of a 'sport.'

The passage of the 1995 St. Croix alewife ban law received little attention -- I did not hear of it until several years later. Few people in Maine outside the local St. Croix River area were even aware of its passage. Nor did they know that by 2000, the St. Croix alewife run had fallen from 3 million fish in 1987 to a just few thousand, and without access to spawning habitat, those few fish left were doomed to extinction.

In 1999 and 2000, the state's fisheries agencies and their Canadian counterparts crafted a compromise plan that would allow for the restoration of an extremely modest and limited alewife run in the river, but to mollify the bass guides, kept alewives out of Spednic Lake. After raucous public hearings in 2000, where the bass guides opposed the plan completely, the agencies persisted in submitting it as an agency bill to the Maine Legislature in the winter of 2001.

Although the bill was specifically designed and 'sold' as a

compromise and peace offering to the bass guides at Spednic, the bass guides didn't bite. Having easily convinced the Legislature to eradicate alewives from the entire St. Croix watershed in 1995, they did not intend to give an inch. But unlike the 1995 law, which was barely noticed in the press, the state's newspapers gave extensive coverage to the 2001 bill. And this is where things really went off the rails.

By 2001 the administration of Maine Gov. Angus King, Jr. was falling into disarray. King, who had been in office since 1994, was a lame duck, prohibited from serving a third term. As usually happens when governors are termed-out, key political appointees start resigning or looking for other employment. Maine DIFW was in such a state. Everybody knew that in January of 2002, a new governor would be elected and new cabinet officers would be appointed. So despite that the 2001 bill to modify the alewife ban had the full support of Maine DIFW at its highest echelons, the DIFW fisheries biologists in eastern Maine, who opposed any alewives in 'their' lakes, actively lobbied through the press during the 2001 legislative debate. The result was not pretty. It revealed for the first time in public the enormous rift within DIFW which had been simmering under the surface for years with its sister agency, the Maine Dept. of Marine Resources.

As 'insiders' knew, there were significant number of DIFW fisheries biologists who were fundamentally opposed to the restoration of any species of sea-run fish to rivers and lakes which they had been allowed by adverse possession to consider 'theirs alone' for the past century. Like the bass guides, some in the field ranks of DIFW saw themselves with the upper hand and were determined not just to preserve their 'gains,' but to greatly expand them. Their public message became that native alewives *anywhere* in the freshwaters of Maine are just plain bad.

Ron Brokaw, a senior fisheries biologist in Downeast Maine, led the charge. In a 2001 *Maine Sunday Telegram* story, Brokaw dismissed the relevance of a joint nine-year study performed by Maine DMR and Maine DIFW which showed no negative interactions between native alewives and bass, trout, smelt and other fish at Lake George, a fairly small natural lake in Canaan, Maine. Brokaw said the Lake George study was irrelevant to Spednic Lake or anywhere else because the study was only done in one lake in central Maine.

Greg Burr, a Maine DIFW fisheries technician, told the press that the native landlocked salmon fishery in Big Lake in the St. Croix had been 'hitting on all cylinders' since alewives began to be kept out of the lake. Rick Jordan, another senior Maine DIFW fisheries biologist, told the press that eastern Maine lakes with both landlocked salmon and native alewives were 'problem' lakes which rarely produced good-sized salmon. [12] Jordan's clear implication was that the native alewives were the 'problem.'

Mike Smith, another senior Maine DIFW fisheries biologist, dismissed any impact on smallmouth bass from the massive 1980s drawdowns by G-P at Spednic Lake, and instead said that alewives were the sole cause of the lake's 'problem.'

To add to the soup, they and the bass guides pointed the press to rare instances where native alewives sometimes passed along a pathogen (erythrocetic necrosis) to rainbow smelt. They also pointed to the Great Lakes, where introduced non-native landlocked alewives created a thiamine deficiency in the (also) non-native Pacific salmon which ate them. It got to the point where if someone dropped an egg on their kitchen floor in Alexander, Maine a biologist in DIFW would put in a call to the *Bangor Daily News* to blame it on an alewife.

As a newspaper reporter in Maine for 15 years, I could see the crazy logic of the biologists' approach. To get to the bottom of each of these scientific claims, all presented as indisputable scientific *facks*, would take even a skilled and dogged reporter many hours of deep scientific

12 The St. Croix is perhaps the worst place on Earth to claim that native alewives are harmful to native rainbow smelt and native freshwater ('landlocked') Atlantic salmon, since all three are native to the St. Croix and have peaceably lived alongside each other since the close of the most recent (Wisconsinan) Ice Age.

literature research to confirm or deny. The citation of impacts by non-native, landlocked alewives in the Great Lakes was a classic 'straw man,' since sea-run alewives are native to Maine, but Maine's newspaper reporters were not astute enough to recognize this clear difference, which any fish biologist would quickly notice.

They Were Never Here in the First Place

The idea of re-writing history was not confined to these eastern Maine DIFW fisheries biologists. The St. Croix smallmouth bass guides began saying that alewives had never lived in the St. Croix River in the first place. They supplied as proof a claim that the lowermost dam on the river, at Milltown, was built on a bedrock falls, and that until the ledges were blasted to build the dam in the early 1900s, the ledges were impassable to sea-run fish. Therefore, they argued, alewives were never even native to the Saint Croix River watershed.

When I first read this claim in the newspaper, I was doing research in the Maine State Archives and came across a bunch of Legislative petitions written in the 1820s by the first white settlers of the St. Croix, complaining that newly built dams on the river were blocking alewives, shad and salmon from migrating 80 to 100 miles upriver. Obviously the bass anglers were factually wrong -- but this did not perturb the Maine press, who dutifully printed the claims of the bass guides and never checked to see if they had any historic validity.

What the two principal reporters (Roberta Scruggs of the *Maine Sunday Telegram* and Diane Graettinger of the *Bangor Daily News*) failed to note in their coverage was that the 2001 agency bill kept in place the total ban on alewives to Spednic Lake, which is where the *faux* controversy had begun 15 years earlier. The Spednic Lake fishermen they were quoting were vehemently opposing an alewife restoration bill that explicitly stated that no alewives would be allowed into Spednic Lake.

David Tobey of the Grand Lake Stream Guides Association

warned that letting any alewives into West Grand Lake would destroy the game fisheries of the entire watershed. Tobey did not explain how alewives could affect West Grand Lake since the 2001 agency plan specifically called for alewives to remain banned from it. The reporters never asked him this question and simply quoted his claim verbatim.

Most conspicuously missing from press coverage was why Maine DIFW was letting its eastern Maine field staff take potshots in the press against a plan their own Commissioner fully supported. This took on a bizarre aspect in Roberta Scruggs' 2001 *MST* feature story, where Peter Bourque, the head of the Maine DIFW's fisheries division stated of the agency plan, "I'm very confident that at the very conservative level there wouldn't be any problem." Immediately after quoting Peter Bourque, Roberta Scruggs quoted Ron Brokaw saying his boss and his agency were totally wrong.

DIFW fisheries biologist Mike Smith, quoted extensively in the 2001 *MST* story, took the discussion to a new level, implying that the entire behavior of anadromy, of fish going back and forth from the sea to freshwater, was bad for freshwater ecosystems. "They'd [the alewives] consume the food that was in Spednic Lake ... It was never recycled back into the lake for other fish to use," he told Scruggs, adding, "Wherever alewives are introduced, they are going to take something away from what's already there."

How did this disconnect occur within the professional aquatic biological community in Maine? A number of deep-seated institutional factors come into play, most of which can be sourced to the splitting of Maine's singular fish and game agency into separate freshwater and marine agencies during the early 1900s. In a state where native sea-run fish originally comprised much of the fish biomass in both coastal and inland waters, this jurisdictional division automatically caused the habitat of sea-run fish to be split down the middle, with their freshwater habitat component under the legal control of Maine DIFW and their marine component under Maine DMR.

The Legislature's early 1900s assignment of sea-run fish

management to Maine DMR set in motion an inevitable schism, since for sea-run fish to exist, they require access to freshwater. In agency parlance, sea-run fish 'belonged' to Maine DMR, but to live they need to spawn in freshwater habitat, which is the sole legal 'turf' of the Maine DIFW. For this reason, many Maine DIFW biologists saw no benefit to them for sea-run fish restoration since the fish ultimately 'belonged' to another state agency, Maine DMR.

For a career inland fisheries biologist like Mike Smith, who focussed exclusively on the needs of sports anglers pursuing purely non-native resident freshwater fish in the lakes and brooks of eastern Maine, it's not hard to see how he would perceive sea-run alewives as 'taking something away' from the St. Croix lakes as they migrated out to sea as juveniles each fall. And for the same reason, a key argument offered by Maine DMR for alewife restoration, that the species is an essential component of the marine food supply in coastal waters, had no relevance to Smith, since the welfare of coastal marine waters was not part of his job description.

This discontinuity has an even deeper taproot; the century-long 'mission creep' of Maine DIFW fisheries biologists from viewing the freshwater environment as a natural ecosystem to viewing it as a disconnected series of waterbodies acting as the vessels for intensively managed gamefish aquaculture with a primary focus on introduced and non-native fish species desired by license-purchasing anglers. On the St. Croix this narrowing of vision put Maine DIFW's fisheries division squarely in opposition with its own wildlife division, who were working diligently to bring bald eagles back to health in Maine.

Maine, like most of the United States, had lost nearly all of its bald eagles by the early 1970s due to DDT spraying. The banning of DDT in conjunction with state and federal laws protecting the bird from hunting and shooting (once a commonplace), the species' elevation to the U.S. Endangered Species list in the early 1970s and protection of its nesting sites had, by the 1980s, allowed for a slow, but modest recovery in Maine. On the St. Croix this recovery was aided and abetted by repairs in fishways on the river beginning in the 1960s and the overhaul of the Milltown Dam fishway near the river's head of tide in 1981 which allowed native alewives to quickly restore themselves in the river and its lakes. This trend was not lost on Maine DIFW wildlife scientists working intensively to restore bald eagles statewide. The shut-off of alewives to the St. Croix, beginning in the late 1980s at Spednic Lake followed by the complete closure of the river in 1995 had a severe effect on the eagle population, as measured each year by Maine DIFW via aerial surveys of active nests and the nesting success.

By 1998, a decade after the initial alewife closure, and the decline of the run from 2.6 million adults to a few thousand, Maine DIFW eagle biologist Charles Todd told a Maine publication, the Working Waterfront News, that the alewife closure was having a disastrous effect on eagle recovery. In the late 1980s, Todd said, he was tracking more than 30 juvenile eagles along the river. In 1998, he said, "There are so few juveniles that we don't even bother to count them anymore." This direct effect of the alewife closure on a signature and endangered wildlife species in Maine and the St. Croix held no sway with Todd's co-workers in the Maine DIFW fisheries division working on the same river. Bald eagle recovery was not part of their job description.

A similar concern was voiced the same year by Edward Baum, senior fisheries scientist with the state's Atlantic Salmon Commission. Baum noted that wiping out the St. Croix alewife run would increase predation pressure by fish-eating birds, such as cormorants, and by smallmouth bass themselves, on the few juvenile Atlantic salmon the St. Croix River still had left in it (in the late 1990s the St. Croix's Atlantic salmon run was numbered in the dozens of adults; today they are extinct). But this factor held little sway with the Maine DIFW's fisheries division, since Atlantic salmon recovery was also not part of their job description. In fact, at this same time, as documents would later show, the eastern Maine DIFW biologists were actively stocking smallmouth bass in native Atlantic salmon watersheds in eastern Maine to create new populations, without even informing their colleagues at the Atlantic

Salmon Authority, who at the same time were trying to rescue the native salmon populations in those watersheds from the brink of extinction. Again, a quirk in state law not only allowed, but encouraged this behavior, since Maine DIFW had sole legal authority to stock any species of fish, in any number it wished, into any freshwater body in Maine without consulting or informing its sister agencies. And while the Legislature gave the Salmon Commission the sole authority to 'manage' native sea-run Atlantic salmon in Maine and sole authority to stock them and regulate their catch (which by 1999 was banned statewide), the Maine Atlantic Salmon Commission had no authority to stop Maine DIFW biologists from stocking non-native, highly competitive species like smallmouth bass directly into Atlantic salmon habitat.

At this point, an outside observer might intelligently ask how Maine's three fisheries and wildlife agencies were allowed to devolve into such an advanced stage of dysfunction, so that one branch of one agency (Maine DIFW-Fisheries) was taking actions that thwarted the efforts of its own wildlife division to restore bald eagles; was stocking an exotic and invasive fish species on top of a nearly extinct native, sea-run species, the Atlantic salmon; and was lobbying for the continued extirpation of another native, sea-run species, the alewife, from its entire habitat in the fourth largest river system in Maine, all to 'protect' an exotic, non-native fish species from a non-existent threat.

The ultimate reason for this self-destructive Balkanization was a failure of leadership from the highest level of the state's executive branch: Maine Governor Angus King, Jr., whose eight year term from 1994 to 2002 encompassed this entire mess. To the extent Harry S. Truman's phrase, 'the buck stops here,' was not rejected wholesale by King, then ultimate blame for this cancerous debacle must be placed quarely on his shoulders. The reason is simple. The 1995 St. Croix alewife bill, which eradicated the largest native population of alewives in the United States, could have been vetoed by Angus King. He didn't veto it.

Prior to the 1990s there was very little published or non-published scientific studies on the interaction of native alewives and other fish in freshwater, including smallmouth or largemouth bass. This was mostly due to the fact that where the species co-mingle, nobody had ever observed any interaction of note, except for the obvious and visible fact that larger bass (ie. three inches or larger) aggressively feed on juvenile alewives.

In the late 1980s the Maine DMR and Maine DIFW and Maine Dept. of Environmental Protection agreed to perform what was hoped to be a definitive study on alewife interactions with other freshwater fish. They chose Lake George in the central Maine town of Canaan to do the study. Lake George is 347 acres, is a natural lake, and is deep enough to create a 'two-story' fish population, with introduced and naturalized trout and rainbow smelt occupying the deeper waters (esp. in summer) and introduced and naturalized smallmouth bass and sunfish in the shallower portions, along with a cadre of native resident fish such as a white suckers, yellow perch and various minnow species. Native alewives had not been present in Lake George since the early 1800s due to dam construction on the Kennebec River.

The idea behind the Lake George study was to intensively examine the stomach contents of resident fish populations for three years, especially rainbow smelt, to determine their diets. Then, for three years, adult Kennebec River alewives were released into pond at a density of six adults per surface acre. The entire study on the resident fish was then repeated with alewives present, along with extensive study of the stomach contents of the juvenile alewives. Then, for the next three years, adult alewives would not be allowed into the lake to spawn and the resident fish population would be studied all over again without the presence of alewives. Water chemistry studies were also done in each of the three, three-year phases.

While the actual field studies were completed on schedule, full analysis of the data and the writing of the report dragged on for many years, in part because Maine DIFW did not allot sufficient staff time to

155

complete their assigned sections of the study. But in 2002 the study and report were finally completed and released. The study found no statistically significant differences in the resident fish population in the three, three-year phases of the study, ie. before alewives, with the alewives, and after the alewives.

This nine-year joint study was done for a very specific purpose: to determine, once and for all, within a rigorous scientific framework everyone agreed upon, if the restoration of native, sea-run alewives to their native habitat in Maine coastal watersheds could have a negative effect on existing, resident fish populations, especially those 'managed' by Maine DIFW for recreational anglers, ie. non-native smallmouth bass, non-native stocked brown trout, native trout (ie. brook trout), non-native freshwater Atlantic salmon, and non-native rainbow smelt. The study results, published jointly by Maine DIFW, Maine DMR and the Maine DEP, found no discernible negative interactions.

In a normal world this would be 'the end of it.' But normal, like natural, is not how things in Maine often operate. During the 2001 Legislative debate, all of the Lake George scientific study data was complete and available to both agencies, even though the final, published report was still in a draft form on a Maine DIFW computer. But, as the press coverage reveals, in 2001 most of the eastern Maine DIFW fisheries biologists treated the Lake George study as if it had never been done, and one biologist, Ron Brokaw, stated it was irrelevant because Lake George is not the same lake as Spednic. To paraphrase George Orwell and Charles Schulz, it was like Lucy pulling the football away from Charlie Brown. Forever.

Now being a cynical sort, I kind of saw this coming, even during the many winters in 1990s when Maine DMR biologist Nate Gray spent most of his work hours dissecting and examining hundreds of smelt and alewife stomachs under a microscope and identifying each pin-head sized creature in them down to the species level. In the back of my mind, I kind of knew that even if the Lake George study showed no negative effects due to the restoration of native alewives, there would be a cadre of Maine DIFW biologists who would choose to ignore the study, because this was never about science in the first place. It was always about turf and who gets to call the shots on 'their' lake.

This is what I told Naomi Shalit, an ex-journalist and recently hired executive director of the fledgling non-profit group Maine Rivers. In the early 2000s, she told me they were going to hire a fisheries scientist from the University of Southern Maine named Theo Willis to do an intensive study on alewife-smallmouth bass interactions with a focus on the St. Croix River and eastern Maine. Knock yourself out, I told her, but you can stack up scientific studies to the sky and people like Ron Brokaw and Mike Smith and the smallmouth guides are just going to say they don't believe them. But Naomi and Maine Rivers did it anyways. The study and paper was completed in 2006 and is quite thorough and definitive.

Willis' study has a number of features that the Lake George study lacked, namely that it examined ten different lakes in eastern Maine, some of which contain smallmouth bass and native alewives, and some, such as the St. Croix lakes, which had smallmouth but no alewives, due to the 1995 alewife ban. Willis put forth that if alewives in eastern Maine truly had a negative effect on smallmouth, there should be a statistically significant and observable difference in the character and health of smallmouth bass in lakes with alewives and nearby lakes with no alewives. He also did stomach analyses of young of year smallmouth and young of year alewives to determine if they were competing for the same food types. Willis' study was designed to test each of the hypotheses made by the eastern Maine DIFW biologists and smallmouth bass guides who believed alewives were bad for smallmouth. And without saying so, it was kind of a statement to Maine DIFW that this is exactly the kind of scientific study that Maine DIFW could have done, and should have done, in the last 15 years if they really wanted to know if alewives had a negative effect on smallmouth bass in the St. Croix and eastern Maine. Much of the historic field data he used was collected by the same Maine DIFW biologists who at the same time were loudly claiming in the press that alewives were bad for smallmouth bass and all other fish. Willis could find no evidence of any negative effects. He wrote:

(1) There is no evidence from available historic data in Downeast Maine lakes that the presence of alewives has systematically harmed smallmouth bass in terms of length, condition or growth. The data provide some evidence that bass grew faster in the presence of anadromous alewives than they did in their absence, though these correlational data do not demonstrate a causal relationship.

(2a) Fish constituted only a tiny proportion (less than 0.15%) of the diet of adult anadromous alewives. Alewives were not significant predators on smallmouth bass, although they did feed on other organisms while present in the lakes. This observation was in contrast to literature assertions that anadromous alewives do not feed while spawning.

(2b) In most lakes, young-of-year smallmouth bass and young-of-year alewives did not have an ecologically significant overlap in diet. In the one lake in which diets were similar, populations of bass and alewives have coexisted for over a century. Based on one year's data, therefore, competition for food between the two species did not appear to be important. Given high lake-to-lake and year-to-year variation in ecological conditions, however, additional data would be welcomed.

(3) Smallmouth bass tournament returns in the past few years have been similar in lakes with and lakes without alewives, suggesting that the quality of sport fishing for bass does not differ systematically between lakes with and without anadromous alewives.

If you're keeping score at home, by 2006 two separate definitive and rigorous scientific studies had been done to test the hypothesis that native alewives negatively affect smallmouth bass in Maine. One study, which took nine years to complete, was done jointly by Maine DIFW and Maine DMR. The second study relied on Maine DIFW field data and closely examined the lakes in and around the St. Croix. Each tested the exact hypotheses put forward by those who said alewives were bad for smallmouth. Both found no evidence to support the claim -- even on the St. Croix itself. And on top of this, not a single study exists anywhere in the world which has ever shown a negative effect by native alewives on smallmouth bass. So that would seem to settle it. Right?

When all of these study data were presented by Maine DMR and Maine DIFW and others to the Maine Legislature in 2007, in yet another attempt to allow for a very small and carefully controlled population of native alewives to be restored to the St. Croix, the Maine Legislature again rejected it, with the one proviso that alewives would be allowed to go into the impoundment of the Woodland Dam, but no further. This restored alewives to none of their native habitat, since they need to spawn in natural ponds, not dam impoundments on rivers, and gave them access to a tiny sliver of water that all biologists considered very poor spawning and growing habitat. The 2008 bill, hailed as an historic 'compromise,' did nothing in real life, since the Canadians had been transporting alewives from the Milltown Dam to the Woodland Dam impoundment since 2001. The only real effect of the 2008 law was that Canadians were spared the annual expense of trucking the alewives over the Woodland Dam.

After the 2008 bill was passed, Maine Rivers and several other Maine conservation groups decided in desperation to appeal directly to the International Joint Commission (IJC), a body created by a 1915 treaty between the United States and Canada to resolve disputes between the two nations regarding rivers, like the St. Croix, which form the international boundary between the U.S. and Canada. The IJC could have, and should have, adjudicated this international dispute in 1987, when Maine first unilaterally shut off alewife passage on the St. Croix at the Grand Falls Dam, but they did not. Nor did they in 1995, or 2001, or 2008. They only entered into the dispute via a pointed request in 2009. In response, they appointed a sub-committee of a sub-committee to investigate the issue. The sub-committee produced a report and recommendations of its sub-sub-committee in the spring of 2010. This report was released for public hearing and comment in the summer of 2010. So what did it say?

While the 18 pages of text, charts, graphs and computer

'simulations' certainly looked 'sciencey,' missing was any discussion of the core scientific question: does any scientific evidence exist to show a negative interaction between native migratory alewives and smallmouth bass? The only reference to the controversy which was the sole impetus for the 'plan' was a brief passage at page 2, which stated: "By the early 1980s improved fish passage and water quality in the St. Croix system resulted in an increasing alewife population. This was *perceived* to have contributed to declining numbers of juvenile bass and poor quality smallmouth bass angling in Spednic Lake." (emphasis added).

Despite the lack of any scientific evidence of a negative interaction between alewives and smallmouth bass, the scientists who wrote the report assumed as fact that alewives do negatively affect smallmouth bass, stating, "To address the concern that the bass reproductive failures in Spednic Lake were directly linked to alewife presence, the plan uses relative young of year abundance [of bass] at the end of their first summer as the measure to ensure bass are not negatively affected by the re-introduction of alewife."

The most troubling aspect of the 'plan' was quietly buried in the charts and graphs: it called for the permanent and total closure of 72 percent of the native alewife habitat in the St. Croix to any alewives, including Spednic Lake, where the whole controversy began 25 years earlier.

Reaction to the IJC 'plan' was swift, voluminous and not complimentary. Jake Kritzer, Ph.D., senior marine scientist with the Environmental Defense Fund stated:

"The plan is not only based on misaligned priorities, but also on a weak scientific foundation. The evidence for a negative effect of alewives on the smallmouth bass population is either non-existent or at best equivocal. Indeed, the plan offers no evidence for a negative impact, and seems to be based entirely on a 'perceived' impact." Yet, alewife and smallmouth bass have co-existed at high abundance for decades after the introduction of the latter in the 1800s. In fact, the more likely scenario is that alewives have the potential to enhance the bass population ... The dual functions of [alewives] of an anadromous fish and a forage fish make alewife a keystone species, the restoration of which should be a top priority."

The National Marine Fisheries Service stated:

"NMFS cannot support agreements that would maintain fish passage barriers to historic spawning and rearing habitat for native sea-run species. Spednic Lake and West Grand Lake and areas upstream of those lakes are not being considered for free access by native sea-run fish such as river herring. These areas represent tens of thousands of acres of suitable spawning and rearing habitat for river herring. NMFS fully supports accelerated and unimpeded recovery of river herring through complete, safe and timely passage at all anthropogenic barriers in the St. Croix watershed. We believe that securing passage prior to the 2011 run is an essential first step to recovery of this depleted species."

The U.S. Fish & Wildlife Service stated:

"Since 1995 the Service has supported the restoration of native diadromous fish including alewife, blueback herring and American eel to the [St. Croix] watershed. Restoring these species to historic habitat in the Gulf of Maine is a priority for the Service. Providing unrestricted free passage alewife to the St. Croix River watershed will contribute significantly toward this goal ...To contribute most significantly to our alewife restoration goals, the entire run should be passed throughout the watershed in perpetuity beginning in Spring 2011."

The Atlantic States Marine Fisheries Commission stated:

"ASMFC believes the proposed plan should be modified to allow for unrestricted alewife access to their historic habitat throughout the St. Croix watershed. The IJC should allow free access to alewives in the St. Croix to gain greatest ecological and economic benefits for both the watershed and regional fisheries."

Aside from these broadsides, perhaps the most trenchant commentary was offered by attorney Sean Mahoney of Conservation Law Foundation (CLF), who noted that the changes in Maine law necessary to implement the 'plan' would violate the U.S. Clean Water Act.

This was a subject that I had been researching independently of CLF in the summer of 2010, when the IJC plan was first made public. By the recent Maine and U.S. Supreme Court decisions on the Presumpscot River, the 1995 and 2008 alewife ban laws by Maine Legislature were flatly illegal under the U.S. Clean Water Act, but nobody had ever challenged them on this basis. The issue came down to a simple question: by what legal authority did the Maine Legislature ban native alewives from the St. Croix in the first place? The text of the laws themselves offered no insight, nor did the transcripts of the legislative debate. Everyone just assumed the Legislature had the authority. But did they? Do they?

Under the Clean Water Act, all states are required to enact water quality standards for all the waterbodies in their state and submit them to the US Environmental Protection Agency for approval. Maine's water quality standards state that all waterbodies must provide 'suitable habitat' for all fish species native to that waterbody. Until the early 1990s, the phrase 'suitable habitat' was loosely defined by state agencies and generally taken to mean only that the water was clean enough to support native fish life in the sense of it not being so polluted by wastewater discharges that no fish could live there.

But in 2003, the state was forced to re-think this interpretation due to a legal challenge by the South African-based SAPPI corporation during the re-licensing of its hydroelectric dams on the Presumpscot River. There, the state opined that SAPPI had to provide passage for alewives, shad, salmon and eels at its five dams to ensure the Presumpscot River above the dams met its water quality standard of being 'suitable habitat' for its native, sea-run fish species. In turn, SAPPI argued that as long as the water above its dams was clean enough to support these fish it was irrelevant whether the fish could actually gain access to the water past their dams 100 year-old dams.. In its order denying SAPPI's appeal, the state wrote:

"Nowhere, as appellant suggests, does the statute state that 'some' of the waters be suitable for the designated uses; that 'some' of the aquatic species indigenous to the waters be supported; or that 'some' of the habitat must be unimpaired or natural. On the contrary the terms 'receiving waters' and 'habitat' are unqualified and the statute specifically states that the water quality must be such to support 'all' indigenous aquatic species ... Appellant's contention that water quality standards are being attained as long as the designated uses of fish, fishing and aquatic habitat are present to any degree in any portion of the river is thus contrary to the language of the statute and to the Legislature's stated objective 'to restore and maintain the chemical, physical and biological integrity of the State's waters.' 38 MRSA Section 464(1)."

In its 2005 decision in *S.D. Warren v. Maine Board of Environmental Protection*, the Maine Supreme Court affirmed the state's reasoning, writing:

"Maine's law is settled in this area. In Bangor Hydro-Electric Co., 595 A.2d at 442 n.4, we concluded that narrative criteria at 38 M.R.S.A. §465 (2001 & Supp. 2004), which requires waters "of sufficient quality to support all indigenous fish species," was intended to be an integral part of the water quality standards for the BEP to consider. We also concluded, based upon the specificity of the designated uses at 38 M.R.S.A. §465, that the Legislature's purpose for the language "suitable for the designated uses" was "that the designated uses actually be present."

In a classic case of 'be careful what you ask for,' Maine's executive branch had convinced its highest court to rule that passage for native, sea-run fish is integral to the state's water quality standards and the U.S. Clean Water Act. In a sense, the Maine DEP had unwittingly sought

163

and received a court decision which invalidated the Legislature's alewife ban on the St. Croix. So the question became, if it's a violation of water quality standards for a private dam owner to block alewives, how is it legal for the Legislature to do the same thing?

In the spring of 2011, the national conservation group EarthJustice decided to take the case, with Friends of Merrymeeting Bay as principal plaintiff, claiming the alewife ban laws unlawfully interfere with the goals of the U.S. Clean Water Act and must be struck down. In spring 2011 U.S. District Court Justice Nancy Torreson dismissed the case, ruling that EarthJustice filed its case under the wrong statute and against the wrong party. EarthJustice, she said, should have filed its case under the U.S. Clean Water Act citizens suit provision and named the US EPA as defendant for its failure to review the St. Croix alewife 'ban' law for compliance with the U.S. Clean Water Act. In May 2012, CLF, EarthJustice and Friends of Merrymeeting Bay sued the US EPA per Justice Torreson's recommendation. On July 9. 2012 the US EPA Region 1 in Boston, Mass. issued a letter order to the Attorney General of Maine, William Schneider, stating that the US EPA found the St. Croix alewife 'ban' law to be an illegal amendment of the state's water quality standards for the St. Croix River and that under the U.S. Clean Water Act the law is null and void. Several weeks later Maine Attorney General William Schneider informed US EPA that the State of Maine would ignore their ruling and would continue to keep native alewives from living in the St. Croix River.

Documentary Records in Chronological Order: 1605-1972.

• 1605 ACCOUNT OF ALEWIVES IN EASTERN MAINE.

Samuel Champlain, describing his 1605 visit to the Schoodic (Saint Croix) River in eastern Maine.

"In May and June so great is the catch here of herring and bass that vessels could be loaded with them ... the Indians resort thither sometimes five or six weeks during the fishing season."

• 1622 ACCOUNT OF ALEWIVES IN PLYMOUTH, MASS.

John Pory, describing alewives going up Town Brook to the Billington Sea (a freshwater pond) in Plymouth, Mass. in 1622.

"In April and May come up another kind of fish which they call herring or old wives in infinite schools, into a small river running under town, and into a great pond or lake of a mile broad, where they cast their spawn, the water of the said river being in many places not above half a foot deep.

Yea, when a heap of stones is reared up against them a foot high above the water, they leap and tumble over and will not be beaten back with crudgels."

From: Pory, John. 1622. Letter of John Pory to the Earl of Southhampton. In: <u>Three Visitors to Early Plymouth</u>. Reprinted by Plimoth Plantation. Plymouth, Mass.

• 1674 ACCOUNT OF ALEWIVES IN MAINE.

"The Alewife is like a herrin, but has a bigger bellie therefore called an Alewife, they come in the end of April into fresh Rivers and Ponds; there hath been taken in two hours by two men without any Weyre at all, saving a few stones to stop the passage of the River, above ten thousand."

From: John Josellyn, <u>Colonial Traveler. A Critical Edition of Two Travels to New England</u> (publ. 1674). Paul J. Lindholt, editor. University Press of New England. 1988.

• IMPORTANCE OF ALEWIVES IN 17th CENTURY NEW ENGLAND.

From the History of Taunton, Massachusetts.

"This is the document which has come into our hands, through the kindness of Mr. James M. Cushman, a direct descendant of Elder Cushman, of Plymouth, and for some years clerk of the City of Taunton -- a document signed by William Briggs Jr. of Taunton, considerably less than a century after the settlement, and who must, therefore have known and conversed with some of the settlers and got his information from them. His father, William Briggs, grand senior (as he designated himself), was a man of substance and good standing in town, as was also the son. The document, in part, is as follows:

'The Indian name for Taunton is Cohannit, at first given to the falls in ye Mill River where the old Mill (so called) now stands, being the most convenient place for catching alewives of any in those parts. The ancient standers remember that hundreds of Indians would come from Mount Hope and other places every year in April, with great dancings and shoutings to catch fish at Cohannit and set up theyr tents about that place until the season for catching alewives was past and would load their backs with burdens of fish & load ye canoes to carry home for their supply for the rest of the year and a great part of the support of ye natives was from the alewives.

"The first English planters in Taunton found great relief from this sort of fish, both for food & raysing of corne and prized them so highly that they took care that when Goodman Linkon first craved leave to set up a grist mill at that place, a town vote should be passed that the fish should not be stopped. It is well known how much other Towns are advantaged by this sort of fish. Middleboro will not permit any dam for any sort of mills to be made across their river to stop the course of fish nor would they part with the privilege of the fish if any would give them a thousand pounds and wonder at ye neighboring town of Taunton, that suffer themselves to be deprived of so great a privilege.

"It seems to be a sort of fish appropriated by Divine Providence to Americans and most plentifully afforded to them so that remote towns as far as Dunstable (as we hear) have barreld y'm up and preserved them all winter for their reliefe. No wonder then that the poor people of Taunton were so much concerned when such sort of a dam was made at Cohannit that should quite stop the fish from going up the river and therefore prosecuted the man that did it in ye law (which process in law how it came to a full stop as it did is mysterious and unaccountable) and it was difficult to persuade the aggrieved people to forbear using violence to open a passage for ye fish and to keep in the path of law for y'r reliefe.

"It is very strange and matter for lamentation that those who complain'd for want of fish were so much derided and scoff'd at as contemptible persons. Strange that any of mankind should slight & despise such a noble and bountiful gift of Heaven as the plenty of this sort of fish afforded to Americans for their support; nay, 'tis very sinful that instead of rendering thanks to our Maker and Preserver for the good gift of his Providence for our support, that wee should despise them. Be sure, many, who formerly saw not that stopping the fish would be so great a damage to the Publick are now fully satisfied that it is an hundred pound damage in one year to Taunton to be deprived of these fish & as the town increases in number of people, the want of them will be found & perceived more and more every year.

"These fish may be catcht by the hands of children in theyr nets while the parents have y'r hands full of work in the busy time of Spring to prepare for planting. Some of Taunton have been forced to buy Indian corn every year since the fish were stopped, who while they fisht, they'r ground used to have plenty of corne for y'r family & some to spare to others. The cry of the poor every year for want of the fish in Taunton is enough to move the bowels of compassion in any man, that hath not an heart of stone."

Source: Taunton (Mass.) Historical Society

• IMPORTANCE OF ALEWIVES TO CORN CULTIVATION -- 1706.

From minutes of Town Meeting, Middleborough, Massachusetts, March 29th, 1706, regarding the town's alewife fishery at Chesemuttock, Nemasket River:

"It is voted that if there be any man in the town that doth not plant any Indian corn, he shall have no turn of fish, and he that plants so little that he needeth not a whole load of fish for it, he shall have no more than for what he doth plant; in which proportion it is to be understood that he shall use but one fish to a hill."

Source: Weston, Thomas. 1906. History of the Town of Middleborough, Massachusetts. Houghton and Mifflin. Boston, Massachusetts.

• KENNEBEC INDIANS CATCHING ALEWIVES -- 1722.

Father Rasle, in the Jesuit Relations, describes in a 1722 letter to his brother annual trips made by the Kennebec Indians at Norridgewock to catch alewives. While the catching site is not specified it is probably either the Sandy River at Farmington Falls or the Sebasticook River at Benton Falls. His description of "large herring" suggests the catch may have also included American shad.

"At a certain season, our people go to a river not very far distant, where during one month the fish ascend the river in so great numbers that a man could fill fifty thousand barrels with them in a day, if he could be equal to that work. These fish are a sort of large herring, very agreeable to the taste when they are fresh; they crowd upon each other to the depth of a foot, and are drawn up as you would draw water. The Savages put them to dry for eight

168

or ten days, and they live upon them during the whole time while they are planting their fields."

• FIRST LAW IN NEW ENGLAND TO PROTECT ALEWIVES -- 1735.

Laws of Massachusetts Bay Colony
Session of the Great and General Court
for 1735-1736

"Chapter 21

"An Act to Prevent the Destruction of the Fish called Alewives.

"Notwithstanding the provision of law already made for removing incumbrances obstructing the natural or usual course of fish, in their season, in brooks and rivers, yet no sufficient remedy is provided where such obstruction is occasioned by dams erected for mills, &c. which is to the grievous damage of his Majesty's good subjects in diverse parts of this province, more especially where such dams have been made across rivers through which alewives or other fish have been wont to pass, in great plenty, into ponds, there to cast their spawns; wherefore, to prevent the like inconvenience and damage for the future --

"Be it enacted by His Excellency the Governour, Council and Representatives in General Court assembled, and by the authority of the same,

"Sect. 1. That no dam shall, hereafter, be erected across any river or stream, thro' which alewives or other fish have been accustomed to pass into ponds, in which there is not made and left a convenient sluice or passage for such fish, on penalty that the owner or owners of such dam shall, upon conviction of failure or neglect therein, before any court proper to try the same, forfeit and pay the sum of fifty pounds; and if the owner of such dam shall not keep such sluice open during the space of thirty days in a year, at least, at such time or times as the alewives usually pass such stream, that then he or they shall forfeit and pay the sum of twenty shillings per day for every day of the aforementioned and limited time it shall not be kept open ...."

Source: <u>Massachusetts Laws, Acts and Resolves</u>. Available at Maine Legislative

Law Library. Maine Capitol Building. Augusta, Maine.

• KENNEBEC RIVER BLACK BEARS EAT ALEWIVES -- 1760s.

"The Worromontogus has one branch -- Meadow Brook, -- which rises in Chelsea Meadow, and has a very considerable fall and mill privilege at the outlet, and after running about a half mile, empties into the main river. The main branch rises in Togus Pond, in Augusta, and runs entirely through Chelsea, and about two miles in Pittston and empties into the Kennebec, being about seven miles long. The water power here is excellent.

"It is related that alewives were so plentiful there at the time the country was settled, that bears, and later swine, fed on them in the water. They were crowded ashore by the thousands. Mrs. David Philbrook, who was a McCausland, was very much in want of a spinning wheel. One day she took a dip net, and caught seven barrels of alewives in the Togus, and took two barrels in a canoe, and paddled them down to Mr. Winslow's, and exchanged them for a wheel."

Source: Hanson, J.W. 1852. History of Gardiner and Pittston. William Palmer, Publisher. Gardiner, Maine.

• FIRST WHITE SETTLER OF PITTSFIELD, MAINE EATS ALEWIVES.

Source: Cook, Sanger Mills. 1966. Pittsfield on the Sebasticook. Furbush Roberts Printing Company. Bangor, Maine.

"Lovel Fairbrother came to the Kennebec at an early day and explored this river and the Sebasticook; found choice intervale at or near the fork of the river, and abundance of fish in the river and game in the forest. He therefore pitched his tent a big camp near the forks of the river in 1775 and moved his family there being joined by two others and this commenced the settlement in what is now the prosperous town of Pittsfield, then called Sebasticook.

"Soon after he got his family there, he was visited by the Plymouth Patent surveyor, who was surprised to find a man of his intelligence in that secluded place to which there was no road; separated from all other settlements by ponds and swamps and impenetrable forests and he took from his haversack a bottle of rum and instated him as Governor of Sebasticook and treated him and he was then called Governor as long as he lived.

"The Governor was disappointed in his expectations. He did not enjoy living upon herring and coarse bread made of pounded corn. There being no mills within 20 miles and no road or communication with other places but by water in the summer and ice in the winter. The land being on Plymouth Patent he could get no title to it; and could have a deed of a lot given to him if would settle in Norridgewock. He in 1777 transferred his possession at that place to Moses Martin who moved there from Norridgewock with his family and spent his days there to old age."

• HISTORY OF ALEWIVES IN THE SEBASTICOOK RIVER, MAINE.

Source: Fisher, Carleton Edward. 1970. History of Clinton, Maine. Kennebec Journal Press. Augusta, Maine.

"For the early pioneers food in the form of fish could be easily had, as there were plenty in the clear, cold waters of the Kennebec and Sebasticook Rivers. During the early period fish were chiefly of value as a food to sustain them, but it was not long before the fishing industry became an important source of income.

"When Rev. Paul Coffin toured the area in July 1796 he reported in his journal:

'July 30th, Clinton. Rode two miles to Capt. Jonathan Philbrick's on Sebasticook, just above the falls, where they catch herring and shad. Thousands of barrels of herring have been taken this spring. They put four quarts of salt to a barrel of them, and when salted enough, they smoke them. They are then handy and quite palatable. Mr. Hudson had three thousand of them hanging over one's head in his shop or smoke house. A pretty sight.'

"George Sullivan Heald described the fishing activities of his father, Capt. Timothy Heald. Captain Heald was living on the Sebasticook in Winslow, but his activities will give some indication of the fishing industry in the area. During 1797 he had a fish seine catching shad and alewives, for which he received one thousand dollars besides some material for building a house. The fish were transported to market in a large box made by laying a double floor of boards twenty feet square, placing boards around the outside until it would hold forty barrels, then the top was covered with two thicknesses and the corners bound. These fish were sold for one dollar per barrel and sent to the West Indies for the Negroes.

"Alewives, also called herring, and shad were the predominant fish to be caught, but some salmon were to be had. The Sebasticook River had fewer salmon in comparison to the Kennebec River. This situation may have been caused by the lack of adequate spawning grounds. In any case, they were not in sufficient quantity to be important commercially, but some of them must have been of good size. Isaiah Brown, who had a store at what is now Benton Station, credited Joseph Proctor for a salmon caught in 1807. Brown wrote in his ledge, "one small salmon, Wt. 7 1/2 lb. at 5 cents per lb., 38 cents." The fishermen in town today would certainly like to catch some of those 'small' salmon out of the Sebasticook.

"Dams, which were so necessary if the mills were to use the water power, did not help the fishing. The first dam, erected at the upper falls in what is now Benton Falls, was built before the Revolutionary War and had a gap for fish. In 1809 another dam, twelve feet high, was built at the lower falls, with no fishway. It stood for five or six years, and in that time had so impoverished the fisheries that the selectmen cut it away. The town in 1814 obtained an act authorizing them to control the fisheries. At the annual town meeting in March 1815, the fish committee was authorized to deliver gratis to each of the town's inhabitants a quantity of fish not exceeding two hundred to each individual. Furthermore, should anyone omit to apply in the season of taking the fish, he was to be entitled to as many from the treasury of the fishery as would be equal in value to the quantity he was to have received from the committee.

"In 1817 it was voted to auction off the fish privilege. The first division, from the Winslow line to Sebasticook Bridge, went to William Richardson, Jr. for $70. The next division, from Sebasticook Bridge to Isaac Spencer's south line, also to Richardson for $117. The third division, from Spencer's line to Capt. Andrew Richardson's south line, went to Joseph P. Piper for $55.50. From Richardson's line to the upper limits of the town,

David Gray paid $16.50.

"While the inhabitants seem to have found it better fishing in the Sebasticook rather than the Kennebec River, this may have been due to two factors: first, the river could be spanned easily by weirs and, second, the town was astride the river. Thus, the voters could control the fishing industry. This was not possible on the Kennebec, for Fairfield had possession of the west bank.

"The fishing soon started to decline. In April 1817 the town voted to petition the legislature to pass laws for the removal of numerous large weirs and other obstructions in the Kennebec River, which were ruining the fishing up the river and on streams emptying into it. Nothing came of this effort. In 1818 the town entered the price-fixing stage in the fishing trade, voting the price of alewives to be two shillings per hundred and shad at six cents each. In 1819 the price of shad was fixed at eight cents.

"In 1838 the last fish treasurer was elected and, although the town voted the following year to auction off the fishing interest, the end to great fishing had come. Its doom had been sealed by the construction of a dam at Augusta; no provision was made for the passage of fish over the dam."

• ALEWIVES AND THE AMERICAN REVOLUTION -- 1776

[Note: The British naval blockade of the New England coast during the Revolutionary War shut off the supply of cod and other ocean fish to Maine's coastal towns. Many towns responded by demanding legal action against mill dam owners who violated colonial law by blocking the runs of salmon, shad and alewife runs in their local rivers.]

Petition of Citizens of Winthrop, Maine -- June 29, 1776

"To the Honorable the Council for the Colony of Massachusetts Bay and the Honorable House of Representatives of the Same in General Court Assembled, The Petition of Joseph Baker, Ransford Smith and Daniel Dudley a Committee of the town of Winthrop in the County of Lincoln in Said Colony in behalf of the Town Humbly Sheweth:

"That Said Town is Situated in the River Called Cobiseconte formerly

173

noted for one of the best streams in these parts for Fishing but some years ago Doct. Silvester Gardiner late of Boston Erected a mill dam at the mouth of Said River where it empties into the River Kennebeck which entirely stopped the Course of the fish up Said River called Cobiseconte. The Inhabitants of Said Town Sensible of the Great advantage of the fish taken so near as they might if they were not stopped by Said mill dam applied to Said Doct. Silvester Gardiner to make a fish way through or round his mill dam which he seemed willing at first to do but after delaying from one time to another refused to do anything about it and the Town having no other way to obtain a course for the fish up Said river but pursuing the measures printed out by the Law of the land which they have been prevented from taking advantage of by the breaking out of the present Troubles and Considering the advantage the fish would be in case they could have a Course up not only to the Inhabitants of Winthrop but to others in the Neighborhood Your Petitioners pray your Honours to take their Case under Your Consideration and Grant Relief by ordering the occupiers of the saw mill dam to make a course for the fish by said dam or otherwise as your Honours in your Wisdom shall See fit and your Petitioners shall ever pray.

June 29 A.D. 1776

Joseph Baker
Ransford Smith
Daniel Dudley
Committee of Winthrop"

Source: Massachusetts Archives on microfilm at the Maine State Library, Augusta, Maine.

Petition of Citizens of Cape Elizabeth, Windham, Gorham and Pearsontown, Maine -- August 22, 1776.

"To the Honourable Council and House of Representatives of the Colony of the Massachusetts Bay in New England in General Court Assembled:

The Petition of the Towns of Cape Elizabeth, Windham, Gorham and Pearsontown in the County of Cumberland.

Humbly Shew:

That the said Towns lay bordering on Presumscutt River, so called, and for many years after the Settlement of this Eastern Country were plentifully supply'd with Salmon, Alewives, Shad & other Sorts of Fish that frequented the said River in great abundance it being peculiarly commodious for the Spawn & Increase of Fish by reason of a large pond called Sebago or Sebacook which extends upwards of thirty miles from the mouth of said River as far as Pondicherry as also the many branches of said River that used to bring a plenty of the aforesaid Fish near to many of our doors, your Petitioners further shew that by reason of several Mill Dams being built quite across the said River, without leaving a sluice way for Fish to pass up, as by Law is directed, and since the said Mill Dams have been erected on the said River the passage of all kinds of Fish as aforesaid has been totally obstructed & stopt in their course up said River to the great prejudice of many back Towns which depended (in their Inland state) on the said River for a part of their support, as Also to the prejudice of all the Inhabitants for the Sea Coast near the mouth of said River by causing a scarcity of Codfish, Haddock, and many kinds of Fish that frequent the mouths of such extensive Rivers after a Quantity of small Bait that abound in such places. And our fishing on the Banks as well as on our Coast off Shore being in a great measure impracticable by reason of the Enemy's cruisers that infest our Coast, reduces us to the necessity of Adopting some method whereby the Fish may come to us. And the Laws of this Colony have been found ineffectual hitherto for the removal of your Petitioners cause of Complaint, Wherefore your Petitioners pray Your Honours to take the matter of our Complaint into your consideration and Grant to your petitioners such relief as in Your great Wisdom & Clemency You may Judge meet & Your Petitioners as in Duty bound shall every pray.

Gorham. August 22nd 1776

George Strout, Harry Dyer
Committee of Cape Elizabeth

William Elder, Zerubebell Hunewell, Thomas Trott
Windham Committee

William Gorham, Prince Davis, Caleb Chase
Committee of Gorham

Daniel Cram, John Deane, Ephraim Rowe
Committee of Pearsontown"

Source: Massachusetts Archives.

• SAMUEL ADAMS SUPPORTS ALEWIVES -- 1785

"Commonwealth of Massachusetts.

In the Year of our Lord, 1785.

"An Act for opening Sluice-ways in the mill-dam or dams which have or may be erected on Presumpscot River, in the County of Cumberland, and upon any Stream or Streams which fall into same river.

"WHEREAS it appears to this Court that the people who live in the neighborhood of Presumpscot River in the County of Cumberland have heretofore, and still may, derive extensive benefits from the fishery on the said river and streams which fall into the same, unless prevented by the mill-dams which have or may be erected across the said river and streams, the increase or even continuance of which unregulated, for any considerable length of time, must inevitably destroy the annual course of the fish up said river.

"Therefore be it enacted by the Senate and House of Representatives in General Court assembled, and by the Authority of the same, That the Court of General Sessions of the peace for the said county of Cumberland, be, and they are hereby authorized and directed, annually to appoint a committee, consisting of three indifferent and discreet persons within the same county, whose duty it shall be to take effectual care that sufficient sluice-ways be annually opened in all mill dams erected, or that may be erected across the said River or Streams, in order that the fish may not be obstructed in their passage up the same, and that the said sluice-ways be annually kept open during the season in which Salmon, Shad, and Alewives usually pass up the said River; which committee so appointed shall be sworn to the faithful discharge of the duties assigned them by this act, before they proceed to the execution of the same duties.

"And it be further Enacted by the authority foresaid, That where the owner or owners of any such mill or mills shall neglect or refuse to open or continue open any such sluice-way or ways in their mill dams respectively, in every such case the said committee, or any two of them, are hereby authorized and empowered to cause the same to be done as speedily as may be; and the owner or owners so neglecting or refusing, upon notice given to them or any of them by the said committee or any two of them for that purpose, shall forfeit and pay a sum equal to the reasonable expence of opening and continuing open any such sluice-way or ways, with the addition of fifty percent. Thereto, which forfeiture shall be recovered by the said committee by action of the case to be by them instituted and pursued to final judgment and execution in their capacity foresaid.

"And it is further Enacted by the authority aforesaid, That so much of the monies recovered from time to time as will be sufficient to defray the necessary expences of opening and continuing open as aforesaid the same sluice-ways, shall by said committee be applied to the purpose, and the overplus accruing by such forfeitures, the said committee shall be accountable for to the several incorporated towns herein mentioned.

"And it is further Enacted by the authority aforesaid, That the said committee shall have such reasonable compensation made them from time to time, for their expences and services arising and performed pursuant to this act, by the several towns now incorporated or may be incorporated, in equal proportion, as do or shall stand in the last preceeding state tax-act, and which towns adjoin the same River, as the said Court may think it proper to allow; and that if any of the said incorporated towns shall neglect or refuse to pay their proportion of the sums that may be due to the said committee from time to time, for their expences and services aforesaid, in every such case, the same committee be, if they see fit, to recover by legal process the whole sum that may be due to them from any one of the said towns which shall so neglect or refuse.

"In the House of Representatives, March 14, 1785.
This bill having had three several readings, passed to be Enacted.
Samuel A. Otis Speaker.
In Senate, March 14, 1785.

This bill having had two several readings, passed to be Enacted.
Samuel Adams, President."

Source: Massachusetts Laws, Acts and Resolves

• DESCRIPTION OF PENOBSCOT RIVER FISHERIES -- 1790.

Statement of Capt. Jacob Holyoke of Brewer, Maine. Born March 27, 1785 in Brewer. Died in Brewer, May 2, 1865.

"I was born March 27, 1785, in the town of Brewer, my parents were living at that time in a log house near the small school house, just above John Holyoke's brick house, where the old cellar hole may now be seen ....

"Mr. John Emory lived at Robinson's cove, about one mile down river; Henry Kenney and John Tibbetts the only other settlers between our house and Col. Brewer's. There were no settlers back and no roads leading back from the river ....

"For many years the Indians were in the habit of making a camping ground of the flat between our house and the meeting house, near the present ship yard, every summer, in going to and returning from the seaboard, where they principally went after porpoises and seals. I have seen often thirty or forty wig-wams, built principally of birch bark, inhabited by two or three hundred Indians.

"There was a beautiful spring of water on the bank of the river, now covered up by John Holyoke's wharf, which the Indians used, and was also used by us.

"This flat of one or two acres was cleared, when my father first came to Brewer, and from the number of Indian stone implements found there in improving the land, was doubtless a very ancient Indian camping ground. When my father built his framed house he cleared up about six acres around it, and upon every side except the river it was a thick, heavy forest.

"Salmon, shad and alewives were very plenty, and in their season many people came here to catch them -- bass were also plenty, and in the fishing season, we could fill a batteau with fish at Treat's falls in a short time; we would sometimes take forty salmon in a day, and I think as many as five hundred were taken some days, in all. My father had a large seine in the eddy, just above the Bangor bridge, and we had much trouble with the sturgeon. When a large sturgeon was captured, the boys used to tie the painter of the boat to his tail and giving him eight or ten feet length of rope, let him go, and when he grew tired or lazy would poke him up with long sticks and so be carried all around the harbor.

"(Signed) Jacob Holyoke. Brewer, Dec. 1860."

Source: <u>The Centennial Celebration of the Settlement of Bangor, September 30, 1869</u>. Published by Direction of the Committee of Arrangements. Benjamin A. Burr, Printer. Bangor, Maine.

• DAM OWNERS PROTEST ALEWIFE PROTECTION LAWS, GARDINER, MAINE -- 1791

[Note: Eighteenth century laws requiring fish passage at mill dams were not popular with most mill dam owners and compliance with these laws was rare. In 1791, Robert Hallowell, Jr., the son-in-law of the founder of Gardiner, Maine, devised a novel argument to avoid complying with newly enacted and much stricter fish passage laws for Maine and Massachusetts -- he denied that alewives, shad and salmon ever went up Cobbosseecontee Stream before dams were built to block them. Hallowell's assertion was quickly rebutted by the citizens of the upriver town of Winthrop, who produced sworn depositions asserting alewives and shad did ascend Cobbosseecontee Stream prior to dam construction.]

"To the Honorable Senate, and the Hon. House of Representatives.

The Petition of Robt. Hallowell, Guardian to Robt. Hallowell Jr.

Humbly Shews

"That upon the River Cobbiseconte in Pittson in the County of Lincoln, two Mills are erected the property of the said Robert Jr. one of which has stood nearly thirty years, and the other about eighteen or twenty years, to the Great Convenience and advantage of the inhabitants of said Pittston, and the Circumjacent Country -- That in order to supply the aforesaid Mills with a Sufficiency of Water, two dams were made and have been continued on said Cobbiseconte river ever since the said Mills were respectively erected, without which the water would be entirely diverted from said Mills and the same would become useless, as to the great damage of the Public, as well as to the said Robert --

"That by an Act made the 29th day of February in the year of our Lord 1789, intitled an Act to prevent the destruction and to regulate the Catching of the fish called Salmon, Shad and Alewives in the rivers and streams in the Counties of Cumberland, and Lincoln, and to repeal all laws heretofore made for that purpose. An authority is given to certain Committees described in said Act, to open and destroy said Dams for the purpose of making a fish way, whereby said Dams are continually exposed to be thrown open & rendered useless. That in the Event no advantage would result to the Community, as the expence of making a fish way would be very considerable, and the same would be in a great measure ineffectual when built, as the oldest inhabitants in that Country cannot recollect any instance of the Alewives proceeding above the aforesaid Dams, and as a variety of natural obstructions render it highly improbable, that the larger fish would ever proceed above said dams in any considerable number --

"Your petitioner therefore prays this Honorable Court to take these facts into consideration, and to appoint a Committee to inspect the premises, so that if it should appear that the damage arising to the community from the destruction of said Mills would exceed the benefit, accruing from the opening a fish way, such measures may be adopted, as will prevent the operation of the Act upon the Dams erected over the aforesaid stream -- Or if in the opinion of the said Committee a fish way should be found expedient, they may in that case be instructed to report, the dimensions, and restrictions under which it shall be made --

Robt. Hallowell."

Deposition of John Stain -- 1790

"I John Stain of Lawful age testify and Say that about thirty years ago before there was any mill Dam built across Cobesecontee Stream I caught Shad fish in said Stream up at the falls about a mile from the mouth of said Stream where a saw mill now Stands and have for years together when I was there to Catch fish Seen Shad and Elwives to go over the falls going up said Stream. -- John Stain. Lincoln, December ye 31st 1790

The above named John Stain made oath that the above Declaration by him Subscribed was true before -- Robert Page, Justice of the Peace."

Deposition of Abraham Wyman -- 1791

"Abraham Wyman of Wyman's Plantation in the County of Lincoln, Gentleman of Lawful age, testifieth and saith that some years before there was any mills built on Cobesecontee stream so called which Emptyes in to Kennebeck River at Pittstown, I was hunting on said Stream and I saw a plenty of alewives Runing up said Stream they were then a mile above what was called the upper falls and further the Deponent saith not. -- Abram Wyman"

Deposition of Joseph Greeley -- 1791

"The Deposition of Joseph Greeley of Sandey river in the County of Lincoln yeoman of Lawfull age testifieth and saith that about four or five and twenty years ago and to the best of my Remembrance it was the year that Cobboseecontee mill Dam was Caried away I was a hunting on Cobbosseecontee Stream so called that Emptied into Kennebec River at Pittstown and up said Stream at the falls in Winthrop where John Chandler Mills now Stand I Saw a Plenty of Alewives Runing up Said falls. I also Saw Major Heald the same day he informed me that he had also Seen them as well as myself and further the deponant Saith not.

Joseph Grele. Lincoln, January 21st 1791

Personally appeared the above named Abraham Wyman and Joseph Grele and after being Duly Cautioned and Examined made Solom Oath to the truth of the Above depositions by them Subscribed before me. Obadiah Williams, Justice of the Peace."

Source: Baxter, James P., editor. 1910. <u>Documentary History of the State of Maine Containing the Baxter Manuscripts</u>. Vol. 22. Maine Historical Society.

Lefavor-Tower Company. Portland, Maine.

• COMMERCIAL RIVER FISHERIES, BANGOR, MAINE -- 1806

Statement of Joseph Carr, Esq. of Bangor:

"In the year 1806 my father built a wooden store now standing on Washington Street at the City Point, between the brick stores built by Zadoc French and Joseph Leavitt, and the wharf known as 'Carr's wharf,' which was the first wharf built on the Penobscot River. In this store my father traded until about the year 1842. All sorts of goods were kept for sale, and Saturday was the great day of trade, and Saturday afternoon (my just holiday) was usually spent by me on compulsion in waiting on my father's customers. On this day there came to the store men from celebrated families of Harthorns, McPhetres, Spencers and Inmans, bringing with them shingles, salmon, shad, smoked alewives and credit, for which they wanted tea, tobacco, calico and rum. It was one if not my chief duty to quench the thirst of these most thirsty customers. Innumerable gills, pints and quarts of good old 'Santa Cruz' have I drawn and delivered to these genial souls, of whom I can truly say none were drunk, but 'all had a drappie in their' ee.' I have now in my possession the original copper gill cup, which furnished those hardy pioneers what they considered to be almost their 'meat and clothing' and their drink it certainly was.

"Santa Cruz rum was one dollar a gallon; New England rum two shillings and sixpence; tea was four shillings and sixpence per pound; tobacco one shilling and sixpence; seven yards of calico made a dress for any ordinary sized woman; salmon sold for four pence halfpenny each; shad and alewives a cent apiece in small lots, or fifty cents a hundred by the quantity; but these last had no pecuniary value so far as a dozen went for any one's individual consumption. I have often seen nets drawn full of shad and alewives in Kenduskeag Stream, both above and below the bridge, and before any wharves were built in the stream."

Source: <u>The Centennial Celebration of the Settlement of Bangor, September 30, 1869</u>. Published by Direction of the Committee of Arrangements. Benjamin A. Burr, Printer. Bangor, Maine.

• PROTECTING ALEWIVES IN VASSALBORO, MAINE -- 1807.

"An act to regulate the taking of fish called Alewives, in a part of Kennebeck River."

"Whereas, the fish called Alewives, are greatly impeded in their passage up Seven Mile Brook, in the town of Vassalborough, by means of seins drawn at the mouth of said brook, in Kennebeck River:

"Be it enacted by the Senate and House of Representatives, in General Court assembled, and by the authority of the same, That from and after the passing of this act, if any person shall by means of seins, or in any other manner take any of the said fish called Alewives, in the river Kennebeck, at the mouth of Seven Mile Brook, in the town of Vassalborough, or within ten rods above, or sixty rods below the mouth of said Seven Mile Brook, at any time in any week, except between the sunrise on Monday, and sunrise on Wednesday in each week; the person so offending, shall forfeit and pay the sum of ten dollars, for each and every time they shall draw a sein within the limits aforesaid, on the days hereby prohibited; and one cent for each of said fish taken in any other manner, to be recovered by the treasurer of said town, and to the use of the inhabitants of said town of Vassalborough, in an action of debt in any court proper to try the same. [This act passed February 25, 1807]."

Source: Maine Laws, Acts and Resolves.

• PENOBSCOT RIVER COMMERCIAL FISHERIES -- EARLY 1800s.

Source: Atkins, C.G. and N.W. Foster. 1869. Commissioners of Fisheries, Second Report. Augusta, Maine.

"The time that we were able to give to the Penobscot this year was mostly occupied in a tour through the fishing district, during the month of May. The weirs were then in full operation and much valuable information was elicited.

"In old times the most abundant fish (in bulk) in this river was the shad; this was probably the most valuable. Next came the salmon. Alewives were exceedingly abundant but little esteemed. Bass (Roccus lineatus, Gill.) were not rare. At Oldtown falls as many shad and alewives were taken as would supply the demand, and many fold more might have been taken; the price, one dollar per hundred for shad, was not sufficient inducement to provide beforehand the necessary barrels and salt to take care of them.

"On the lower part of the river the market was more convenient, many vessels, mostly from Connecticut, coming every season to load with shad and salmon. Immense quantities of them were shipped in this way. Before the river was closed with the dams the price of salmon had risen to six cents a pound, that of shad to six cents apiece. Alewives, smoked hard for the West India market, brought in early times thirty-three cents a hundred in Boston, and the price afterwards rose to one dollar and one dollar and quarter, when they were very profitable. The fishing, previous to 1785, was all done with nets, but they have been gradually superceded by weirs and at the present time very few nets are used. Their use, however, was continued as long as it was profitable. At one time there were, it is estimated, two hundred men employed in drifting between Mill Creek and Odom's Ledge."

• PETITION OF THE PENOBSCOT INDIANS, 1821

"Petition of the Chiefs of the Penobscot Tribe of Indians praying a law may be passed to prevent the destruction of fish in the Penobscot River.

"To the Whole Legislature of the State of Maine

"We the undersigned Chiefs & others of the Penobscot Tribe of Indians ask you to hear us in our petition in which we mean to speak nothing but truth and first we would say that in the days of our forefathers the great plenty of fish which yearly came into the waters of our Penobscot River was one of the greatest sources by which they obtained their living and has so continued within the remembrance of many of us who are now living which plenty we always considered as sent us by the Great God who provides means for all his Children --

"But when our white brethren came amongst us they settled on our

184

lands at and near the tide waters of our River and there was plenty of fish for us all -- but within a few years our brethren the white men who live near the tide waters of our River have every year built so many weares that they have caught and killed so many of the fish that there is hardly any comes up the River where we live so that we cannot catch enough for the use of our families even in the season of the year when Fish used to be most plenty.

"We have asked the general Court at Boston to make laws to stop the white people from building weares and they have made Laws but they have done us no good for the Fish grow more scarce every year. Besides the weares they use a great many long nets. We can only use very small nets and spears -- now we ask you to make a Law to stop the white folks from building any more weares forever so that Fish may again become plenty and also stop the white people from using any seines above Kenduskeag on the main river.

"And we ask you to make the Law so as to stop the white people and Indians from catching fish more than two days in the week in the season of Salmon, Shad and Alewives at least for five years. We think that Fish will then be plenty again.

We are your Brothers.

John Neptune, Chief

Source: Maine State Archives. Augusta, Maine.

• PETITION OF PENOBSCOT BAY WEIR FISHERMEN -- 1821

[This mass-produced petition is one of the earliest fishing industry lobbying efforts in Maine and illustrates the size of the Penobscot River commercial fishery during the early 1800s. The use of "half tide" weirs on the Penobscot River and Bay began in 1813 for salmon, shad, alewives and striped bass. This new technology resulted in an 'arms race' between Penobscot River drift netters and weir fishermen. Each group deployed bigger gear each year and blamed the other for the continuing decline of the fish.]

185

"To the Senators and Representatives of the State of Maine, in General Court convened.

GENTLEMEN -- The undersigned, residing near the waters of Penobscot bay and river, respectfully represent,

"That the fisheries of said bay and river have, for several years, been shackled with so much restriction and regulation, that our rights to fish at all are nearly annihilated. The liberty to fish, commences under many grievous conditions on the 20th of May and continues to the 4th of July. That memorable day, on which our liberties begun, ends the rights of the fishermen. No matter whether owing to a late or forward season, the fish take it into their heads to run a month later or earlier as frequently proves to be the case, the poor fisherman cannot, upon penalty of a great fine, take a scale after the said 4th of July; and even during this scanty period, we are by law, allowed to fish with wears but four days in seven. There would be more reasonableness in this law, if an act of parliament or General Court, could as well restrain and limit the running of the fish, as it does the fishermen; but these lawless depredators upon your flats and shores, laugh at all your fish laws and will not heed your commands as in any limitation of time; your statute books are unknown to them. The right of government to regulate the fisheries (not to destroy them) in the channels and deep waters, is an inherent one; but whether the government has at any time a constitutional right to prevent us your petitioners from building or erecting wharves, booms, or other machinery on our own ground, or on our own flats, admits of a question, which we are confident merits and will receive your most cautious and deliberate investigation. We purchased the soil, "with all the privileges thereunto belonging," and in many instances, gave an exhorbitant price for the very privileges we now contend for. If the government once presumes to dictate to the subject, whether he shall not make improvements on his own estate, although we might apprehend no danger from present members of the Legislature, yet, to what tyranny and despotism might not such a stretch of power lead in the hands of the ambitious and unprincipled? Freemen ought to resist usurpation in its incipient stages. We have never delegated to you the power to enter upon our estates, to build up this fence, or pull down another. We have not yielded to you the right to decide whether we may, or may not, build wharves and wears on our own flats. We have, indeed, by a sort of common and implied consent, given to our Legislature the right to regulate the fisheries, the right to keep open the channels for navigable purposes -- and the general right to do any thing which is for the manifest advantage and benefit of the whole community, not however to the sacrifice and destruction of private property without ample recompence.

"We have looked forward to the organization of our new government with the most pleasing promise of a redress of wrongs. We advocated our independence under the expectation, that the government of Maine would better understand the rights and necessities of its own citizens, than a government whose sun shone upon us at a great distance, whose invigorating influence hardly ever penetrated the coverts of our wilderness country. We approach you with respect, but without humiliation, and resolutely remonstrate against the laws now in force, "regulating the salmon, shad and alewive fisheries of the Penobscot bay and river."

"First Reason: Because we are restricted to about 24 days fishing, after deducting what is called "unlawful days" during the "permitted time" the fish may not happen to run.

"Second Reason: The selectmen of towns are invested with royalty powers, of granting or refusing license to build wears. They are also empowered to exact a bond, with ample sureties, for each fisherman's "good behavior" before any offence has been committed.

"Third Reason: Certain officers and executors of the law are appointed at the discretion of the Selectmen called "fish wardens" who have proved arbitrary in the exercise of their "little brief authority," insolent, vexatious, and unprincipled, with a few exceptions only, so that the very term has become a cant word of reproach and no honest man will, from choice, accept the appointment.

"Fourth Reason: The restrictions on wear-fishing prevent the taking of small fish, on which numberless of the poor depend for subsistence, and on most of the privileges where wears are built, nets cannot be set, on account of the strength of the tide or current.

"Fifth Reason: Contained in the affidavits herewith transmitted.

"On these restrictions and limitations we offer this single comment, that all laws for the government of the people, in order to insure them respect and willing obedience, must be just and reasonable.

"Our "red brethren" have been instigated by some of their white brethren, far up the river, to make a talk about the destruction of salmon, by our expert fishermen on the big waters -- It will be found on investigation, that they have contributed their full share, to the destruction of the fish, not for their own use or consumption, but for fish merchants. When a salmon has run the gauntlet and arrived unharmed at the still waters, where the spawn is deposited, it becomes an object of solicitude; for by spearing them in these retired places, as has been the constant practice of the Indians, the destruction of a single fish is that of thousands. Here it is then, if any where, that arbitrary and insolent fishwards should be appointed to execute the vengeance of the laws. The Indians are now reduced to mere handful of strollers, having no regular residence and have really little or no interest in the result. All of which is submitted for your consideration, with that deference, which is always due from the people, to the impartial and upright Legislature of their choice."

William Wardwell and 175 others.

Source: Maine State Archives. Augusta, Maine.

Note 1: This petition was written after and in response to the Jan. 1821 petition of the Chiefs of the Penobscot Indians asking the Legislature to restrict the weir and driftnet fisheries in the lower Penobscot River and Penobscot Bay. The use of the phrase "our red brethren" in the final paragraph is a sarcastic reference to the Penobscot Chiefs' use of the phrase "our white brethren" in their 1821 petition. Given the sentiments toward the Indians expressed in this petition, it seems doubtful the authors considered the Penobscot Indians to be their brethren.

Note 2: Several of the regulations criticized in the above petition are found in the 1816 Massachusetts Law, Chapter CXLIV, titled "An Act for the preservation of Fish in Penobscot River and Bay, and the streams entering into the same." This Act required all weirs to be removed from the river by July 5th of each year with a fine of $50 for violations. The law authorized fish wardens to remove or demolish any weir left in the river after July 5th and required all weirs to be licensed by the Selectmen of the town adjoining the site of the weir and assessed a fine of $100 for unlicensed weirs.

• PETITION OF INHABITANTS OF PHIPPSBURG, MAINE -- 1821.

[Note: Another petition from weir fishermen, this time on the lower Kennebec River.]

"To the Honorable Senate and House of Representatives of the State of Maine in Legislature assembled

"The subscribers inhabitants of the Town of Phippsburg respectfully ask leave to represent that the Inhabitants of said Town are deeply interested in the Fisheries. First, that a large number depend almost entirely upon the River Fishery. Second, that a still larger proportion of our Inhabitants as well as those of the neighboring towns, and even Fishermen from various other parts of the State are wholly dependent at certain seasons of the year, on the wears erected at and near the Mouth of the Kennebec, for bait fish, in order to pursue the Codfishery; that the owners of these wears are at great expence in erecting these, whereby a large number of poor persons are employed, which is a great means of support in the Spring of the year.

"Your Petitioners further represent that to be deprived of the privilege and means of taking fish called Salmon, Shads and Alewives, a privilege which we and our forefathers have enjoyed of a time immemorial, would not only be depriving your Petitioners of the principal means of support, but would subject many others of our Inhabitants to great distress, and thereby become chargeable to the Public. That to be deprived of the means of obtaining Baitfish, for carrying on the Codfishery would be subjecting a large number of the Inhabitants of our State on the Seaboard to the greatest inconvenience.

"Your Petitioners further represent that in their humble opinion the wears and other obstructions at the Mouth of the Kennebec are not the cause of the Dimunition of fish on said River, the said wears do not in any degree obstruct the fish passing up and down the Channels of the River; that from the outer part of the wears to the shore on the opposite side of the River, the space is no part less than half a mile distance, that the wears are without exception erected on the Flats which we hold by the same right as the lands adjoining, that with two exceptions only, the wears are up the Bays and Coves and quite aside from the main passage of the fish in the River.

"Your Petitioners are of the opinion that the Dimunition of Fish on the Kennebec is caused by the erection of Mill Dams and by other Obstructions on and across the Streams and Brooks, when the fish usually go up to cast the spawn, that many schools of fish, particularly of the Salmon and Alewives which formerly frequented those Streams and Brooks are known to have been entirely cut off by the erection of the Mill Dams and other obstructions which prevented them from going up to cast their spawn. That the Salmon in great numbers formerly passed up the Androscoggin, but since the erection of Mill Dams across said River, a School of Salmon called "The Brunswick School" have been entirely destroyed or left the River, and to prove this and many other important circumstances, the testimony of many aged and respectable Citizens can be produced.

"Your Petitioners further represent that the Lumber towards the mouth of the Kennebec is almost entirely exhausted, that the state of the Soil principally by the Sea Coast is such as to render it impossible to obtain support by Agriculture. Thus your Honours will be convinced that the privilege and means of taking fish at and near the mouth of the Kennebec is of the first and greatest importance to the subsistence of a great number of Citizens. Your Petitioners therefore pray your Honors that they may be continued unmolested in the enjoyment of their rights and privileges they now possess and in duty bound every pray. Phippsburg, Jany. 24, 1821"

Andrew Reed, John Parker, Francis Wyman, Joseph Morse and 38 others.
Source: Maine State Archives. Augusta, Maine.

• PROTECTING ALEWIVES IN THE ST. CROIX RIVER -- 1822.

"To the Honourable Senate & House of Representatives of the State of Maine:

"We the undersigned, citizens of said State, respectfully represent that previous to existing obstructions, by mills and mill dams, on the St. Croix or Schoodic River, great quantities of Salmon, Shad & Alewives annually passed up and returned down said River, to the great benefit and advantage of the community generally; and in an especial manner of the new settlements in the eastern part of the State --

"That said obstructions have rendered it almost impossible for the Shad & Alewives to pass above the Town of Calais; whereas they used to pass from eighty to a hundred miles above; and they are now almost totally excluded from said River --

"That it is confidently believed that if suitable fish ways should be provided & also suitable regulations for the taking of fish on said River, it would, as formerly, be abundantly supplied with fish, and all the privileges and advantages of the proprietors of the mills & mill dams on said River remain unimpaired --

"Wherefore, we pray, that such fish ways and such regulations concerning the taking of fish on so much of said River and its branches as be within this State as may be deemed necessary to restore to its citizens their ancient privileges in this respect, may be provided by the Honourable House of Representatives and as in duty bound we will ever pray.

Joseph Whitney and many others. December, 1822.

Source: Maine State Archives

• PROTECTING ALEWIVES IN GOULDSBOROUGH, MAINE -- 1824.

"To the Honorable Senate and the Honorable House of Representatives of the State of Maine in Legislature assembled.

"The Petition of the subscribers, inhabitants of Gouldsborough in the County of Hancock, humbly represents.

"That the Stream emptying into Prospect Harbour in said town, called Prospect Stream, was formerly visited, in the proper season, by great quantities of Alewives, which used to go up said stream to a pond at the head thereof, and there cast their spawn -- that for a number of years past their passage up said stream has been obstructed by a mill-dam erected near the mouth thereof, so that few if any Alewives now pass up said stream -- that in consequence of the obstruction aforesaid they have now mostly forsaken said Harbour and

Stream; greatly to the injury of the Cod-fishery on the neighboring coasts; as it is well known that the Cod follow the alewives, in great numbers, even into the Bays and Harbours where they frequent -- that a convenient and sufficient passage for said fish may be made through or around said dam at a small expence, and without material injury to
the Mills situated thereon.

They therefore humbly pray your Honours to pass an Act for opening said Stream, and establish such regulations on the subject as wisdoms shall judge proper and expedient. As in duty bound will every pray.

Gouldsborough, Dec. 20th, 1824."

Robert G. Shaw and 35 others
Source: Maine State Archives. Augusta, Maine.

• PETITION OF INHABITANTS OF BURNHAM, MAINE -- 1827.

"To the Honourable Legislature of the State of Maine, January Term, A.D. 1827

"The Inhabitants of the Town of Burnham in the County of Kennebec respectfully represents that formerly the Alewives used to pass up the Stream of the 25 Mile Pond in great abundance and that for several years past there has been a Mill Dam erected across the 25 Mile Stream in the Town of Unity and there has not been a Sufficient Sluiceway through said Dam to permit the alewives to pass up said stream into said pond to cast their natural spawn nor for the fry to pass down said Stream and we would further represent there has been several wares made across said Stream above and below said dam for the purpose of taking said fish at the season of the year when said fish pass up said stream into said pond and they have taken said fish in said wares every day in the week, Sunday not excepted, and it is much doubtful whether any of said fish were permitted to pass into the said pond during last year as there was none seen or known to be in said pond during last season and there was a considerable quantity taken at said wares and we therefore request your Honourable Body to take the subject into consideration and if in your Wisdom you shall think it proper you will pass a Law they there shall be such laws and regulations on said 25 Mile Pond and stream as there is on the Sebasticook River for the preservation of said fish as duty bound will every pray."

Hezekiah Reynolds and numerous others.
Source: Maine State Archives. Augusta, Maine.

• PETITION TO PROTECT ALEWIVES, KENNEBEC RIVER -- 1827.

Petition of Charles Hayden and 52 others -- 1827

"To the Honorable the Legislature of the State of Maine

"We the undersigned inhabitants of several towns in the vicinity of the Kennebec River respectfully represent that the fish called Salmon, Shad and Alewives which pass up the river every spring of the year are considered of great importance not only as a convenience but comfort and help to support many of said inhabitants, and that said fish are greatly obstructed and destroyed in their usual passage up and down said river by reason of numerous machines and obstructions placed in said river to take, kill and distroy said fish.

"At the mouth of said river or near thereunto are placed a multitude of wears for the purpose of taking said fish which prove very destructive by killing and breaking the schools of fish and driving them back into the ocean so that it is believed that comparative few in number make their way up the river. Next they are met in almost every eddy and mouth of small streams by nets of enormous lengths until they arrive at Ticonic falls between Winslow and Waterville, where the fish are met by new extraordinary and sure instruments of death called traps which placed in almost every avenue where it would be possible for the fish to run. These machines, implements or contrivances to take and kill said fish are kept almost constantly in the river have nearly distroyed the whole run of said fish. And at present fish laws for said river, if any there be, is found altogether inadequate for the purpose of protecting the passage of said fish up and down said river.

"We therefore earnestly request that the Legislature at its present session would enact such a law for the whole river Kennebec and Sebasticook as will give said fish a complete protection in their passage up and down the same and that the fish may have at least three days in each week to pass free of any obstruction. That all nets or seines used for the purpose of taking fish

193

should be of limited and proper length and all improper obstructions removed and forbiden for the future and such a fine or penalty imposed as will deter any person from violating the law -- which should be so plain that he who reads may understand, and will give us a complete relief as it respects the above premises and in duty bound will every pray.

January 9, 1827"

Charles Hayden and 52 others.

Source: Maine State Archives. Augusta, Maine.

• NEWS ITEM FROM THE PENOBSCOT RIVER -- 1829.

"A true fish story -- Seven thousand shad and nearly a hundred barrels of alewives were taken in Eddington last week by Luther Eaton, Esq. at one haul -- Bangor Register."

Source: Kennebec Journal, edition of May 26, 1829. Available on microfilm at Maine State Library. Augusta, Maine.

• PROTECTING ALEWIVES IN PERRY, MAINE -- 1829.

An Act to prevent the destruction of Alewives in Little River, in the town of Perry. Laws of the State of Maine. Maine Laws, 1829, Chapter 20.

"Be it enacted by the Senate and House of Representatives, in Legislature assembled, That the owners or occupants of such dam or dams as now are, or hereafter may be erected across Little River, so called, in the town of Perry, in the county of Washington, shall make and maintain a sufficient sluice or fish-way, round, through or over such dam, or dams, for the passage of Alewives, and shall keep the same open and free from all obstructions, from the twentieth day of May to the first day of July each year; and any owner or occupant of such dam, or dams, who shall neglect to make, maintain and keep open such fishway as herein directed, shall forfeit and pay the sum of one hundred dollars, to be recovered by action of debt in any Court of competent jurisdiction, one moiety thereof to the use of said town of Perry, and the other moiety thereof to the use of any person who may sue therefor.

Approved February 13, 1829."
Source: Maine State Archives. Augusta, Maine.

• PROTECTING ALEWIVES IN NEWPORT, MAINE -- 1835.

"To the Members of the Legislature for the State of Maine now in session at Augusta:

"Represents the undersigned individuals residing in the Town of Newport in the County of Penobscot that they have been for a number of years past much obstructed and deprived of the benefits of the alewife fishery in the waters of the stream running through said town called the Eastern branch of the Sebasticook, which said stream has heretofore furnished valuable quantities of fish to the inhabitants of said town.

"That they have been deprived of the benefits of said fishery in consequence of obstructions being placed in said stream and in the main branch of the River Sebasticook by the owners of mill dams and wears without leaving sufficient fish ways through the same. Your Petitioners further represent that a Special Act was passed by the Legislature of A.D. 1826 entitled, "And Act to Prevent the Destruction of Fish in the Sebasticook River" with a view of remedying the evil complained of by your petitioners. But on account of the limited language of said act your petitioners derived no benefit therefrom, the said act extending its remedies only to the main river and not to its branches, upon one of which your petitioners do reside.

"Your petitioners do therefore request that an addition may be made to said act so as to extend the requirements and penalties of said act to all streams and branches of said river (as well as the main river itself) in which the fish called Salmon, Shad and Alewife or either of them have been in the habit of passing up to leave their spawn. And further that an action of debt may be commenced to recover the penalties provided by said act, by an persons of any town through which the said Sebasticook river or any of the aforesaid streams and branches may pass to be appropriated as provided in said act, and before any court of competent jurisdiction and in any county wherein either of the parties in said suit shall reside.

Newport, Jany. 19th, 1835."

Source: Maine State Archives. Augusta, Maine.

• PROTECTING ALEWIVES IN THE ST. CROIX RIVER -- 1836.

"Baring, November 15th, 1836

"To the Honourable Senate and House of Representatives in Legislature Assembled:

"The undersigned inhabitants of the Town of Baring in the County of Washington, respectfully represents that formerly the fish called Salmon, Shad and Alewives were very plenty in the River St. Croix and its Branches --

"That said fish were of great utility to this portion of the community and tended much to promote its settlement --

"That the number of said fish has been rapidly diminishing of late years, owing, principally, to the dams and obstructions that have been built across said River --

"Your petitioners believe that if a law were passed compelling the owners of mills on said river to build suitable fishways round, through or over the mill dams on said river and also regulating the times and days of taking said fish, and requiring the fishways to be kept always open and the wears to be kept shut two days in each week, from the first day in April, to the first day in September, in each year, and prohibiting all persons from taking said fish on said days, either in wears, seines, drift nets, set nets, scoop nets, or with spears, that said fish would soon become plenty in said river and its branches, and greatly tend to promote the interests of this community, and the settlement of the wild lands in this vicinity --

"Your petitioners therefore pray, that your honorable bodies will pass such a law relating to fishways and the taking of said fish in said river, as you, in your wisdom may think best calculated to promote the public good."

Matthew Fowler and others.

Source: Maine State Archives

• PROTECTING ALEWIVES IN THE SANDY RIVER -- 1836.

"To the Legislature of Maine

"We the undersigned citizens of the State respectfully represent that great injury is experienced by the good people of this State and particularly by that portion of them which reside on, and in the vicinity of the waters of the Kennebec River and its branches by the dams which have been erected across the Kennebec and its branches especially that branch called Sandy River and thereby preventing the free egress and regress of those Fish called Alewives, Shad, Salmon and Trout, and indeed for every kind of Fish which formerly passed up the waters of those Streams before the erection of said dams.

"And we further represent that the privilege of these kinds of Fisheries is of great and essential benefit to the public and to be deprived of them is a public injury which in our humble opinion requires redress. And for that purpose we earnestly solicit the attention of the Legislature to the subject and pray that passage ways through the several dams across the Kennebec River and its branches may be kept open at those seasons of the year when, or during which these several kinds of Fish usually pass up these streams."

(Signed) O.L. Currier and numerous others.
Source: Maine State Archives. Augusta, Maine.

• MILL DAM OWNERS PROTEST FISH PASSAGE LAWS, WARREN, MAINE -- 1837

[Note: The decline and extirpation of alewife runs across Maine in the 19th century resulted in a bewildering number of laws to force dam owners to comply with fish passage requirements. In many instances, dam owners protested just as vigorously, devising an argument known today as 'payroll vs. pickerel.' The following exchange of petitions by mill dam owners and citizens on the St. George River in Warren, Maine is typical of the time. The St. George alewife run is one of very few in Maine that was not extirpated by dams and today supports a town alewife harvest in Warren which provides several thousands a year to the town coffers.]

"To the Honorable Senate and House of Representatives in Legislature assembled:

"Your petitioners, inhabitants of the town of Warren, respectfully represent that the law regulating the taking of fish in the St. Georges River in said town requires that the dams across said River shall be opened during the season when the fish pass up & down said River. This is a great injury to the owners of the mills situated on said river and others having business at said mills.

"That the law requiring the dams to be opened aforesaid prevents the vacant water privileges on said stream, which are among the first in the state, being occupied for Factories and various other machinery, depriving the inhabitants of those advantages which Nature has provided them, and thus retarding the growth and prosperity of said Town and the good citizens of the surrounding country.

"That the right of taking fish in said Town or the law regulating the same has become a bone of contention among the people and a prolific source of Litigation some contending that they have a right to take fish in the navigable waters in said Town, and taking them accordingly notwithstanding the law, others owning farms bordering on the river, contending they have a right to fish in the waters upon their own land, and fishing accordingly, the law to the contrary notwithstanding. Thus are generated heart burnings, strife and lawsuits.

"The fish for years back have been much diminished, and they do not when obtained half pay for the trouble and expense of taking and curing them to say nothing of the great waste of time by our citizens in congregating and waiting about the fishways.

"On the whole your petitioners are satisfied that it would be for the benefit of the citizens of this and the neighboring towns to have the law aforesaid repealed, and they do hereby respectfully request the Legislature to repeal the same."

Letter of Town of Warren, Maine in response to the above petition.

"Warren, February 6, 1837

"At a legal meeting of the inhabitants of the town of Warren qualified to vote in town affairs. Voted that our Representative in the Legislature be instructed to oppose any petitions that may be presented to repeal the law regulating the Shad & Alewife fishery in the town of Warren. Voted that our Representative be furnished with a copy of the above vote by the Clerk.

Attest: Stephen Burgess, Town Clerk"

Source: Maine State Archives. Augusta, Maine.

• PENOBSCOT RIVER DAM WARS -- 1838

[Note: During the mid 1830s, two large timber crib dams were constructed across the lower Penobscot River. These were called the "Corporation Dam" at the river's head of tide in Veazie and Eddington, Maine; and the "Great Works Dam" approx. 10 miles upriver in Old Town, Maine. The owners of both dams sought and received corporate charters from the Maine Legislature to build the dams. These Legislative Charters required the corporations to build sufficient fishways at both dams. Fishways were not built. Soon after the dams were completed the Penobscot River's runs of native sturgeon, striped bass, Atlantic salmon, American shad and alewives veered toward extinction; and the lucrative commercial fishery in the river and Penobscot Bay collapsed. According to contemporary accounts, fish wardens attempting to enforce the State's fish passage laws at the dams were threatened with lawsuits by dam owners and in some cases, physically threatened by mill employees. For the next decade the Maine Legislature was showered with citizen petitions arguing for and against the extinction of the Penobscot River's native fish and its commercial fishing industry. Although crude fishways were built at the some of the Penobscot River dams after the Civil War, the Penobscot's commercial fisheries for all species except Atlantic salmon was destroyed by the Civil War and has never recovered. The Penobscot's commercial Atlantic salmon fishery was closed by the Maine Legislature in 1948 and its recreational salmon fishery was closed in 1999. In 2009 the Atlantic salmon of the Penobscot River were declared an Endangered Species by the U. S. Dept. of Interior under the U.S. Endangered Species Act.]

Petition of Inhabitants of Orono and Milford, Maine -- 1838

"Petition of inhabitants of Orono and Milford for repeal of all fish preservation laws on the Penobscot River.

"We the undersigned inhabitants of Orono, Milford and elsewhere on the Penobscot River beg leave to remonstrate against the passage of a bill for an Act entitled, "An act additional to an act for the preservation of Fish on the Penobscot River and Streams tributary thereto" -- copies of which were ordered to be printed and are now before the Legislature. For the following among the many reasons that may be offered why said bill should not be passed.

"First, because it is provided therein that the fish wares constructed in the tide waters and which have ever been the means of obstructing the passage of Fish in said River, are allowed to be maintained without any material curtailment of privilege and which are constructed at short distances from each other and starting from opposite shores so interlock as to render it almost impossible for fish to escape. We do know from experience and observation that whenever these wares are erected and maintained for two or three years in succession (which is even the case when fish are plenty) it has diminished and nearly destroyed the run of fish so as to render the maintenance of wares unproductive and they have consequently with few exceptions been abandoned until by such abandonment (facilitating the passage) the run has become restored.

"Secondly, we object to the passage of the bill because it confers extraordinary powers which are to be vested thereby in men to take away and remove what they may deem obstructing to the passage of fish regardless of the establishment so to be removed and its bearing on the vital interests of the Community and on which supposed obstructions such interests may altogether depend, thereby prostrating enterprise by demolishing establishments on which is founded not only individual interest of great magnitude but the best hopes of the Country and on the very basis of its advancement and prosperity, and which should not we deem be subject to the caprice of any man or number of men having minor and conflicting interests to promote and thus be clothed with impunity.

"We would cheerfully secure if practicable and consistent with interests of far greater moment the unmolested passage of fish in said River. But we cannot but hope the inquiry will be made and duly reflected upon whether the enterprise and resources of the valley of the Penobscot shall be forgone for the sole purpose of securing the privilege of taking for a few days or weeks in the season a small supply of paltry fish which we may venture to say for the fifteen or twenty years past has occupied the Husbandmen about all of that season, which should have been devoted to Agriculture in order to secure a Harvest and have thereby been rather a curse than a blessing to our Country.

"With these suggestions we have the subject to the wisdom of the Legislature to be properly investigated and hope and pray that this contemplated Act be found so apparently obnoxious to the course of improvement on the Penobscot may be refused a passage and that all Laws respecting fish on said River, above Frankfort, may be repealed."

Jesse Wadleigh and numerous others.
Source: Maine State Archives. Augusta, Maine.

Petition of Inhabitants of Orono, Milford and Bangor, Maine -- 1838

"Petition of inhabitants of Orono, Milford and Bangor for repeal of all fish preservation laws on the Penobscot River.

"To the honourable Senate and House of Representatives of the Legislature of Maine,

"The undersigned inhabitants of Orono, Milford and Bangor would respectfully represent that they have seen published an act entitled an act for the preservation of Fish in Penobscot River, now before your honourable bodies and believe that many provisions in said act is predicated upon the principle that the fishing interest is paramount to all others.

"The undersigned believe it would be injurious to the public interest and subversive of private rights to compel mill owners to remove their dams for the preservation of Shad, which are of little importance compared with their interests and the taking of which is now very limited.

"Your memorialists would inquire of your honourable body whether it would be right to subject the mill owners to the sacrifice of their main interest that supports this section of the State in order preserve one that supports none. The undersigned would therefore pray your honourable bodies to repeal all laws respecting Fish on said River above Frankfort believing the public interest requires it."

Salmon Hackett, Jr. and numerous others.
Source: Maine State Archives. Augusta, Maine.

Petition of Inhabitants of Bucksport, Maine -- 1838

"To the Legislature of Maine in session at Augusta, Jan. 1838

"The undersigned citizens of the town of Bucksport in the County of Hancock respectfully represent, that the Act passed in the year 1835 for the preservation of Salmon, Shad and Alewives in Penobscot Bay and River, and their tributary waters, does not accomplish the object for which it was intended. Many Mill Owners on the said waters refuse or neglect to open fish ways through their dams and other obstructions, and threaten the County Fish Wardens with a suit at law if they proceed to open them.

"The said Act requires the ward or wardens to open fish ways through dams and other obstructions (where the owners or occupants refuse or neglect to do it,) but does not sufficiently guarantee them against the strong combination of interest, which would in many instances be brought to bear upon them. Being doubtful how a suit against them might terminate, and not being prepared to encounter Lynch Law, they have neglected to do (as well they might) what one clause of said act makes their imperative duty. Hence the fish in many places are obstructed in their passage up the aforesaid waters, and must soon become extinct unless a radical remedy is provided.

"Believing as we do, that this section of the State could not have been settled and brought forward as it has, had it not been for the primitive blessing of taking fish in Penobscot Bay, river and tributary waters -- and knowing as we do, that there is an ample sufficiency of water in Penobscot river and its tributary streams; for moving all necessary machinery, and for the preservation of fish; we earnestly pray your honorable body, to give the subject your candid and unprejudiced consideration, and pass an Act which shall ensure the preservation of Salmon, Shad and Alewives in the aforesaid waters, as long as 'Old Penobscot rolls his current on.'"

Daniel Buck and numerous others.
Source: Maine State Archives. Augusta, Maine.

Petition of Inhabitants of Orland, Maine -- 1838.

"To the Honorable Legislature of Maine assembled at Augusta, A.D. 1838.

"The undersigned citizens of Orland, County of Hancock, respectfully represent that the fisheries of the Penobscot River and its tributary waters have formerly been a great benefit to all the Inhabitants in this section of the State.

"But for some years past the Salmon, Shad and Alewives, which were formerly abundant, have greatly diminished already, and unless a remedy is provided by law the benefit derived from the said fisheries must be entirely lost -- and many poor people in consequence be deprived of a great part of the means for their support.

"We therefore pray your Honorable Bodies to give this subject (which to the people in this vicinity is of vast importance) your impartial consideration and pass a law which will preserve Salmon, Shad and Alewives in the said River and tributary waters.

Orland Jany. 15, A.D. 1838"

Asa Torrey and numerous others.
Source: Maine State Archives. Augusta, Maine.

Petition of James Austin and others for fishways on the Penobscot River -- 1839

"Petition by James Austin and 107 others requesting that sluice ways be opened for fish on the dams on the Penobscot River.

"To the Honorable Senate and House of Representatives in Legislature Assembled for 1839:

"We the undersigned respectfully represent to your Honorable bodies that it is necessary and desirable for the preservation of fish in the Penobscot River and its tributary streams that some immediate action should be had upon said waters for the purpose of making suitable and convenient passage ways over and through the mill dams now erected on said waters, as many of said Dams are so erected as to almost entirely preclude the passage of fish up the said Penobscot River, and in fact on many of its important Streams the fish are now entirely shut out, and should the said waters remain in their present situation without further interference by your Honorable bodies in a few years the fish in the Penobscot River would become extinct to the great detriment of the many and for the advantage of a few --

"As your Honorable bodies may be well aware some corporations have been granted which have already proved disastrous and detrimental to the community at large, in obstructing the navigation of said River as well as preventing the passage of fish, the fish taken from the waters of said River have to our knowledge for many years proved an advantage to a very great portion of the community by means of which the poor have been supplied and the hungry fed, a blessing provided by nature and which we wish to have remain -- but under present circumstances about to be wrested from the many and that too as we believe without their consent for benefit of the few --

"Gentlemen, we wish to call your attention to corporations and associated wealth with their onward march, their derogation from justice, and their encroachments upon the rights of others, and that too almost without remedy, we can rely upon your Honors alone for protection, and to you Gentlemen we do most pathetically appeal and we do trust we shall not appeal in vain --

"It may be brought against this petition that provision has already been made to open Dams for the passage of fish, we can say with the utmost confidence that so far as provisions have been made they have been almost totally disregarded and to this particular we wish to call to your attention, and as it regards the last Act passed for the preservation of fish in aforesaid waters, a number of the most important Streams were exempted by law and we believe, Gentlemen, unnecessarily exempted too -- we do believe that Cold Stream with some others ought not to be exempted and that measures should be adopted such as you may deem necessary for carrying provisions into effect after provisions are made -- (we now understand that a Dam across the Penobscot River at Old Town is in contemplation and that too is contemplated by a charter from the Legislature) should an exemption be made to carry this measure into effect it is one we most sincerely deplore as it would add another obstruction to those already in existence, believing that you now have the fullest sense of the community on irresponsible corporations and that at least in this State they will henceforward be watched and guarded with a jealous eye, and that you will with the same watchfulness guard the rights of of individuals from the onward march of associated wealth and corporations which are already heavy upon us -- and as in duty bound would ever pray."

James Austin and 107 others.
Source: Maine State Archives. Augusta, Maine.

Petition of Ware Eddy and Others for Repeal of all Penobscot Fish Laws -- 1843

"To the Honorable Legislature of the State of Maine

"Respectfully represent, We the undersigned, citizens of Towns bordering on the Penobscot River, that owing to the obstructions in said river Fish that have been wont to pass up and breed in said river have greatly diminished. A few Salmon and Shad only pass up during the spring freshets and Alewives are hardly seen at all. The many Mills that have been erected and the dams necessary to keep the Mills in operation together with the vast amount of logs and other timber driven down the river in the season most propitious for the passage of fish up have tended to destroy many of the fish or drive them into other waters.

"And as it is not to be presumed that the lumber operations on said river will be suspended, or even restricted, on account of the lesser interest (fishing), the expediency of continuing the present fish law is rendered more than doubtful, or indeed of making any law for the preservation of fish upon said River or the waters thereof. The Law is onerous, expensive and altogether inefficient. It creates officers whose salaries are paid by towns where no adequate service is rendered more to the towns than the public at large. It increases our taxes without an equivalent and it does not effect any good purpose. Where lumbering is carried on to the extent that it is on this River fish are driven off and all the Laws in the world will not bring them back unless the greater interest is subservient to the less. We therefore pray you to take this subject into consideration and inquire into the expediency of repealing all fish laws operating or designed for the Penobscot River or its tributary waters, which will relieve them from the expense we are now subjected to and from officers we have no voice in making. And as in duty bound pray."

Ware Eddy and numerous others. Nov. 30, 1843.
Source: Maine State Archives. Augusta, Maine.

Petition of Citizens of Orrington, Maine - 1843

"To the Honourable Senate and House of Representatives of the State of Maine in Legislature assembled

Orrington, January 1843

"We respectfully represent that in former years before the Corporation and Great Works Dams were erected across the Penobscot River there were large quantities of Shad and Alewives taken in and above the tidewaters of the said Penobscot; that the quantity taken in the tidewaters alone (by a careful investigation) is found to have been in a single year when marketed to amount to more than one hundred fifty thousand dollars and more than four fifths of said sum was the product of labour and that immediately after the erection of said dams the quantity of fish taken rapidly diminished and for the last few years have become almost extinct -- the quantity taken for market for the last year or two has been less than ten thousand dollars and not one fourth part used for home consumption that was formerly used, thus we are very materially injured by said obstructions and are deprived of the just rewards of our labours.

"We further represent that the present laws for the preservation of Fish are wholly inoperative and inefficient. We believe some persons that are acquainted with the fishing business should be delegated and authorized to see that there is good and sufficient fish ways and that the owners or occupants of said dams should be requested to make and keep open the same subject to proper and suitable penalties neglecting to do so. We therefore pray your Honourable body will make such provisions by law as may be necessary to give said Fish (shad and alewives) a passage up the Penobscot River and its tributaries to cast their spawn that we may again enjoy our rights."

Source: Maine State Archives. Augusta, Maine.

Petition of Bucksport and Orland, Maine -- 1844

"To the Senate and House of Representatives of the State of Maine, at Augusta assembled:

"The subscribers, inhabitants of the towns of Bucksport and Orland, beg leave respectfully to remonstrate against the petitions of 'Ware Eddy & others, praying for the repeal of the fish law passed at last year's session.'

"There are two dams only that obstruct the free passage of salmon, shad and alewives up the Penobscot river, 'The Corporation,' owned by John Otis and others, and the 'Great Works,' owned by Josiah S. Little and others.

"These gentlemen lawyers have had the cunning to evade for years that portion of the law as it stands in the Revised Statutes which requires them to open a fish-way by their dams, and it is found insufficient to effect that purpose.

"The law of the last year will compel these soulless Corporations to open a passage way for said fish, which can be done without danger to those structures, and comparatively speaking, at a small expense.

"Your remonstrants also request that the operation of said law be extended to Eastern river, and its tributary streams. February, 1844"

Source: Maine State Archives. Augusta, Maine.

Petition of the Inhabitants of Frankfort, Maine -- 1844.

"Petition against repeal of the fisheries preservation laws on the Penobscot River

To the Hon. the Senate and House of Representatives in Legislative assembled:

"The undersigned, inhabitants of the Town of Frankfort, respectfully represent, that in consequence of the Mill-dams & other obstructions upon the Penobscot River and its branches, the Salmon, Shad & Alewives, which once abounded in said River & Streams, and which serves for food for the inhabitants, have now nearly left us; and learning that some of our most wealthy men (who have made themselves rich by taking said fish and then erected Mill-dams on the River & Streams, and thereby robbed the poorer inhabitants of their natural rights) have petitioned you to repeal the Fish Law passed last year. We therefore remonstrate against the repeal of the Law, believing that the fish will return to us should that Law be suffered to exist and as in duty bound will ever pray."

Elish Chick, Jr. and 22 others.
Source: Maine State Archives. Augusta, Maine.

• ALEWIVES IN THE CONCORD RIVER, MASSACHUSETTS -- 1846

Source: Thoreau, Henry. A Week on the Concord and Merrimack Rivers, in The Writings of Henry David Thoreau. Houghton Mifflin. Boston, Mass.

"Salmon, shad, and alewives were formerly abundant here, and taken in weirs by the Indians, who taught this method to the whites, by whom they were used as food and as manure, until the dam, and afterward the canal at Billerica, and the factories at Lowell, put an end to their migrations hitherward; though it is thought that a few more enterprising shad may still occasionally be seen in this part of the river. It is said, to account for the destruction of the fishery, that those who at that time represented the interests of the fishermen and the fishes, remembering between what dates they were accustomed to take the grown shad, stipulated, that the dams should be left open for that season only, and the fry, which go down a month later, were consequently stopped and destroyed by myriads. Others say that the fish-ways were not properly constructed. Perchance, after a few thousands of years, if the fishes will be patient, and pass their summers elsewhere, meanwhile, nature will have levelled the Billerica dam, and the Lowell factories, and the Grass-ground River run clear again, to be explored by new migratory shoals, even as far as the Hopkinton pond and Westborough swamp.

"One would like to know more of that race, now extinct, whose seines lie rotting in the garrets of their children, who openly professed the trade of fishermen, and even fed their townsmen creditably, not skulking through the meadows to a rainy afternoon sport. Dim visions we still get of miraculous draughts of fishes, and heaps uncountable by the river-side, from the tales of our seniors sent on horseback in their childhood from the neighboring towns, perched on saddle-bags, with instructions to get the one bag filled with shad, the other with alewives.

"Shad are still taken in the basin of Concord River at Lowell, where they are said to be a month earlier than the Merrimack shad, on account of the warmth of the water. Still patiently, almost pathetically, with instinct not to be discouraged, not to be reasoned with, revisiting their old haunts, as if their stern fates would relent, and still met by the Corporation with its dam. Poor shad! where is thy redress? When Nature gave thee instinct, gave she thee the heart to bear thy fate? Still wandering the sea in thy scaly armor to inquire humbly at the mouths of rivers if man has perchance left them free for thee to enter. By countless shoals loitering uncertain meanwhile, merely stemming the tide there, in danger from sea foes in spite of thy bright armor, awaiting new instructions, until the sands, until the water itself, tell thee if it be so or not. Thus by whole migrating nations, full of instinct, which is thy faith, in this backward spring, turned adrift, and perchance knowest not where men do not dwell, where there are not factories, in these days. Armed with no sword, no electric shock, but mere Shad, armed only with innocence and a just cause, with tender dumb mouth only forward, and scales easy to be detached. I for one am with thee, and who knows what may avail a crow-bar against that Billerica dam?—Not despairing when whole myriads have gone to feed those sea monsters during thy suspense, but still brave, indifferent, on easy fin there, like shad reserved for higher destinies. Willing to be decimated for man's behoof after the spawning season. Away with the superficial and selfish philanthropy of men,—who knows what admirable virtue of fishes may be below low-water-mark, bearing up against a hard destiny, not admired by that fellow-creature who alone can appreciate it! Who hears the fishes when they cry? It will not be forgotten by some memory that we were contemporaries. Thou shalt erelong have thy way up the rivers, up all the rivers of the globe, if I am not mistaken. "

• REPORT OF THE PENOBSCOT RIVER FISH WARDEN -- 1848.

[Note: This document is interesting because of the heartfelt and optimistic outlook of its author toward the recovery of the Penobscot River's fish runs after crude fishways were built on some of the river's dams in the mid-1840s. Subsequent records indicate that while Atlantic salmon regained their foothold in the Penobscot River during the latter half of the 19th century, the river's runs of American shad, alewives, sturgeon and striped bass never recovered.]

"Communication of Benjamin Shaw, Fish Warden, related to fishing on the Penobscot:

G.P. Sewal, Esquire
Old Town, 7th of July, 1848

Dear Sir,

The fish wardens of the Penobscot waters have ascertained that in order to preserve and perpetuate the Salmon, Shad and Elwives in the River that section of the Fish Law that provides for the exemption of Penobscot Bay and several of its tributary streams from the operation of the Law thereby giving exclusive privileges to some which are denied to others. Its unequal operation renders the Law unpopular and the Fishermen are not much disposed to observe the Law from respect to it. Permit me hear to say that Laws suitable to command the respect of New England Live Yankies or Penobscot Fishermen must be tinctured a little with Justice and Equalities to command quiet and ready submission.

I now say repeal that section of the Fish Laws that provides for exemptions of certain places from its operation. This will remove causes of complaint and the River will again Swarm with fish as of Old. I say do this and you give the Fish Law that principle of justice and equality that every American admires. This done, not only the fishermen but the more opulent dealers that furnish amd supply them will cheerfully combine in supporting the Law for their preservation and all concerned will be more likely to treat the authority of the Fish Wardens with more Respect. I have stated my views on the importance of repealing this section of the Law, which is so unequal in its operation thereby removing all cause of discontent. Their seems to be a necessity of doing what I have proposed or abandon protection of Fish.

The run of Fish this present season is so great that the people acquainted with their condition are anxious to give all possible protection. And many now think with proper Management they may be increased nearly to their original condition. This is decidedly my own opinion judging by the abundance now known to be in the River. And already on their spawning ground there has been more Fish taken at the foot of Grand falls this year than at any one place on the River, and what were taken were small compared with the quantitys known to be there. It is supposed by those most experienced in Fishing that there is Salmon and Shad suficient at their journeys end to bring in next year an old fashioned Run provided the Laws are good and Equal and well enforced.

The quantities of Elwives is much less in proportion as they are not as strong and less able to perform the long journey. They frequented the lower branches of the River where their favourite places of resort are now wholly closed and they are shut out. This is the reason of their falling off. I am impatient to have liberty to open Mr. Blackman's Stream in Bradley which should never have been closed; and bid welcome the finney Millions to return as of old to the great joy of themselves and the Inhabitants of the Region roundabout. I am informed that Elwives have been seen lingering round below that dam every year since the Law gave leave to close it against them. A good Fishway may be made cheap and be profitable to the owner and all the Country roundabout.

Two important points more I desire to bring to your notice. On the subject of the Penobscot Fisheries to which I ask your patience and consideration; the three years labour as Fishwarden have convinced me beyond a doubt that good laws suitably enforced may bring back the Fish in all their Original Glory and Grandeur without any Detriment to any other of the great interests of the River, but rather a benefit to all others. One other important point is the very great bodies of water held back for the purpose of driving logs and the addition of the Allagash have now a tendency greatly to improve the river during the journey of the fish, of which nearly or quite balances the obstructions caused by the Mills and lumbering.

Again the Fish and actual Settlers have always gone together. The presence of the Fish always has induced Agriculturists to settle in their neighborhood. If they continue to go as they have this year we may expect many Farmers to turn their attention that way. If the Fish fall off the Farmers clear out South.

I have given you this sketch of my views designed equally for your consideration and Friend Richardson together our Friend Stubs intends being at Augusta toward the close of the Session for the purpose of recommending one section of the Fish Law repealed and I respectfully request you to assist him it being very important. All well good growing weather but little news here.

I remain Gentlemen very respectfully yours,

Benjamin Shaw"

Source: Maine State Archives. August, Maine.

• FIRST LIFE HISTORY SUMMARY OF THE ALEWIFE -- 1867

[Note: In their first report to the Maine Legislature, Maine Fisheries Commissioners Charles Atkins and Nathan Foster provided one of the first scientific summaries of the life history and habits of the alewife in Maine.]

"Though inferior to its elder brother, the shad, both in size and quality, the alewife excels in numbers and hardiness. Vast numbers once swam in all suitable waters through the State; and it is found from the Bay of Miramichi to the Chesapeake. To the north of us it is called "gaspereau." In the Middle States, and in many localities in Maine, it is called "herring." In our own State, it has endured against the disadvantages that man has put in its way much better than the shad or salmon. There is less wildness and timidity about is character than is the case with those fish. It is a domestic sort of fish, taking so kindly to civilization, that it has been the subject of numerous experiments in cultivation, so successful that they will deserve some notion by and by.

"The alewife, in migrations, generally precedes the shad into the rivers; but in Eastern River, Dresden, the shad come earliest. They are taken together by seines and weirs. Yet the alewife often chooses for its spawning grounds quiet lakes and ponds, and to reach them pushes up out of the rivers into the smallest brooks, which the shad never does. It seems particularly to delight in shallow, boggy waters, yet it is capable of breeding in tidal waters, as it does in the Kennebec. Clear, cold streams it always avoids.

"Alewives begin to appear in our rivers in April, sometimes in March. By the first of May a few of them are taken in Dresden, and in Augusta. Yet the main body does not appear until late in May, or, in some rivers, until June. The fishermen distinguish three separate "schools," or "runs," of different sizes, and appearing in succession, the first run being the largest and most valued. Of the first run in East Machias, 370 fill a barrel; of the second run, 400; of the third run, 600. Those of the third run, although small, are yet fat and good.

"Unlike the salmon, alewives are deterred from entering a stream by an unusual flow of water, and always wait until it partially subsides. Their movements are consequently irregular in point of time. They advance by day in all difficult or exposed places, as in the passage of rapids and fishways, falling back or remaining stationary during the night. Warm, sunny days are particularly acceptable to them, and they may then be seen in great multitudes. Although of small size, they will stem very considerable rapids, and reach great altitudes, if at the end of their journey there is a suitable breeding place. Their limit on the Sandy river was 120 miles from the sea; on the east branch of the Penobscot, not much less than 200 miles.

"From Mr. John Brown of Bowdoinham, we have learned the following facts in relation to the spawning of alewives. In the month of June, in shallow water, over weedy flats, and along the edge of the channel, they may be seen and heard rising repeatedly to the surface, making a great swirl in the water and disappearing. On observing closely, it was found that several alewives, sometimes as many as six or eight of both sexes, rose together, and the eggs and milt could be distinctly seen falling to the bottom. To make certain, some of them were caught in the act, and search at low water revealed at a little depth multitudes of eggs among the weeds on the bottom in the same spot where the fish had been observed. The operation is performed oftener at night. It has been accurately observed in a weir, where the eggs dropped upon a board floor. About the middle of June begin to be seen in the water of the bay around Abagadasset point, myriads of pairs of eyes, each pair with a tail. Whether these were shad or alewives the observers were unable to determine, but since the experiments at Holyoke indicated that the young shad seek the centre of the river, it is probable that these were alewives. In the fall they can be distinguished, and many alewives linger there in November.

"After spawning the alewives commence their return to the sea. The time when they reach it varies with the distance they have to travel. In some cases they have pushed up into small ponds or pools, whose overflow is so slight that a few days of dry weather completely dries it up, and cuts off their retreat; in this situation they sometimes have to wait until the fall rains release them. Mr. Treat, in his experiments at Red Beach, found that the old alewives came down early in July, having a very short distance to travel. They were followed by the young late in July and early in August.

"The descent of the young alewives generally occurs later than this -- extending into September. It is a most interesting sight to witness their march. They proceed in dense column, frequently miles in length, following all the sinuosities of the shore. Over falls they let themselves down tail first, as indeed all fish do. If obliged to pass a precipitous fall they are not much injured by it, unless violently thrown against rocks or the apron of a dam. When no other way presents, they will pass through an ordinary mill wheel, apparently with little harm. When so small and light they are much less liable to injury than the full grown fish.

"When the young alewives first go to the sea, they are two to four inches long. How fast they grow from that time is not certainly ascertained, but we have reason to believe that they do not mature in less than three years. It has been generally found that when a piece of water has been newly stocked with alewives they do not reappear until the third year. At Red Beach, Mr. Treat saw nothing of his until the fourth year, when they came to the mouth of the stream in great numbers. That they do not die immediately after spawning, as has by many supposed to be the case with shad, has been abundantly proved. Mr. Treat shut them into one of his ponds and kept them five months; at the end of that time they seemed to be much improved. To ascertain the cause he opened several of them found their stomachs fill with their own young. This sort of cannabilism is, without doubt, exceptional. In their natural condition it is not probable that they feed upon other fish.

"Alewives are neither so timid nor so tender as shad. They can be dipped out of the water and put into tubs without injury, and can by an occasional change of water be carried many miles overland. Advantage has been taken of this, to restock some waters that had been exhausted."

Source: Maine Fisheries Commissioners First Report, 1867. Maine State Archives.

• ALEWIFE RESTORATION SUCCESS IN MAINE -- 1867

From Maine Fisheries Commissioners First Report, 1867.

"Instances of Success

"Now, supposing these conditions all fulfilled, what reason have we to expect success? All the materials for an answer to this question that lie before us are too voluminous to be presented. We can only select. And first let us quote some instances of success at home.

"The East Machias River was originally an excellent alewife river, but by the erection of impassable dams and reckless fishing, they were eighteen years ago reduced to a yield of two barrels yearly. By the construction of fishways and careful attention the yield has now been raised to $1,000 or $1,500 yearly; the price being from one to two dollars a barrel. The cost of the fishways was less than $1,000.

"The Cobscook or Orange River, in Whiting, was practically depopulated by dams, not more than a dozen alewives being taken yearly, and those at the head of tide. In 1861 alewives were carried into the lakes, and fishways built; in 1867 an abundance of fish crowding the fishways.

"Dennys River. Alewives and salmon formerly plenty: but greatly diminished; the alewives being practically exhausted by impassable dams. Obstructions being removed in 1858, the alewives have increased, as witness the number caught as follows: in 1865, 2 bbls.; in 1866, 15 bbls.; in 1867, 240 bbls."

• ALEWIFE RESTORATION PLAN FOR THE SEBASTICOOK RIVER -- 1867.

From Maine Fisheries Commissioners First Report, 1867.

"The Sebasticook is a tributary of the first rank. It is the outlet of many lakes and ponds of which the principal are China lake, Unity lake and Newport lake, having an area of 4,000 acres each. This characteristic rendered it principally an alewife river, and of those fish it produced immense numbers. It also yielded a great many shad, and some salmon. The most fish were taken in the town of Clinton, now Benton, and the town was vested with the right to take the fish by their agents, a fish committee, subject to certain conditions. They were to distribute a certain number gratis to the poor, and then sell to the inhabitants at a set price, and finally could dispose of the residue as they saw fit. Great quantities were sold to strangers, the ordinary price being twenty five cents a hundred. Newport also had full control over the fisheries in that town. There were free fisheries on all other parts of the river and its tributaries. Indeed the fisheries were all free until a falling off in supply warned the people that there must be some regulations. On this point we have the testimony of Mr. Beriah Brown of Benton, now 78 years old. Seventy years ago he followed the man who took the fish. Also of Maj. Japeth Winn, who has lived in Benton fifty-five years. The tributaries of the Sebasticook were very early obstructed by dams through which, in most cases, inefficient fishways were left -- generally a mere gap, or a pile of stones; and the number of the fish had been falling off for many years before the town of Clinton assumed control of its fisheries. The dam at the upper falls in Clinton was built before the war of 1775, but a gap for fish was left in it. About 1809 a dam was built at the lower falls twelve feet high, with no fishway. It stood five or six years, and in that time had so impoverished the fisheries that the selectmen cut it away, and allowed the fish to ascend to their breeding grounds. The town in 1814 obtained the act authorizing them to control the fisheries, and the first year after cutting away this dam the fishery was leased for two or three years to one James Ford, he agreeing to pay yearly 200 fish to each man, woman and child in Clinton, and to sell as many more as should be wanted at a set price. From this time the fish increased again rapidly and the town began to sell the fishery yearly at auction. The price obtained varied from $500 to $1,200 or $1,500; the purchaser being bound to distribute gratis to the poor and sell to all townsmen at a fixed price. The year of the closing of the Augusta dam the fishing sold for $225. One or two years before for $500.

"Mr. John Holbrook, 65 years of age, has lived in Newport all his days.

Within memory alewives came here in great numbers, with a few shad and now and then a salmon. Forty-five years ago they were not so plenty as formerly. Thirty years ago they began to diminish rapidly, and in a few years were entirely gone. The obstructions on the Sebasticook now existing are six dams. The dam at Benton lower falls has a sluiceway twenty feet wide and three feet deep, near its west end, which was not closed during the last season until the 20th of June. With a suitable arrangement of the plank this might answer for the passage of fish. Over the upper dam a way might be easily constructed at the east end by bolting down some timbers and blasting a short passage out of the ledge. At Clinton and Detroit the task would be easy, but they must be guarded against ice. At Newport the milldam would require a fishway, but presents no difficulty. The dam at the outlet hardly hinders the passage of fish. The river was not examined above this point, although the alewives used to run as far as Stetson Pond.

"Of the branches we examined the Pittsfield branch as far as Moose lake or pond, the Twenty-five Mile stream, -- and have gathered some information about others. The west branch from Moose lake has three dams, one at Pittsfield and two at Hartland, neither of which presents any difficulty in constructing fishways; all three would require them. At Hartland there has been a dam 67 years, but as long as the alewives came there was a hole left for them to pass into Moose lake. Into the latter runs Main stream, crossed by several dams which were not examined.

"The Twenty-five Mile stream is the outlet of Unity lake. Near its mouth, in the town of Burnham, is a dam built 35 years ago, 12 feet high. Seven miles up the stream is another dam, and beyond that Unity lake. Tributary to Twenty-five Mile stream is Sandy stream of rapid flow, obstructed by two dams. The streams draining Lovejoy's and Pettie's ponds are obstructed each by one dam. The latter has a dam which has stood without a fishway for 60 years. The stream draining Plymouth pond has four dams. The Vassalborough stream is much obstructed, but was not examined.

"All the lakes and ponds of Sebasticook river are admirably adapted to the breeding of alewives. The restoration of these fish would be a comparatively easy matter. Plenty of the live fish or their spawn can be obtained at Augusta or below. The construction of ten fishways would give them access to the three largest lakes with a surface of 10,000 or 12,000 acres. If undertaken on the right scale and perseveringly carried forward great return might be expected in a few years. Abijah Crosby of Benton, was an enthusiast on this subject; who might have accomplished much had he been supported by public opinion. He went so far as to introduce live alewives to Pettie's pond, Unity and Newport lakes; they bred there, the young fish were seen going down the stream, and some of them caught; fishways were built over several of the dams on the Sebasticook, and thad that built at Augusta proved a success, the alewives would now have been again established in the Sebasticook river."

Source: Maine Fisheries Commissioners First Report, 1867. Maine State Archives.

• FISHWAYS BUILT FOR ALEWIVES ON ST. CROIX RIVER -- 1871.

"I was able, in my last report, to announce the construction of the fishways over the dams at Calais and Baring. I am now able to report the success of the experiment. The fishway in the dam at "Middle Landing" or Union Mills, the first obstruction met by fish in ascending the river, was completed in 1869, and has thus been tested during one season, and through several freshets of unusual violence. When, in the month of June, the alewives came, they readily found the entrance to the fishway, and passed up through it in great numbers. Crowds of people gathered to witness the ascent. "

Source: Maine Fisheries Commissioners Report for 1871. Maine State Archives.

• IMPORTANCE OF ALEWIVES TO THE COD FISHERY -- 1872.

Letter by Spencer F. Baird,
U.S. Commissioner of Fish and Fisheries

Washington, D.C.
November 16, 1872

To E.M. Stillwell, Esq., Bangor, Maine

My Dear Sir,

I am in receipt of your letter, asking my opinion as to the probable cause of the rapid dimunition of the supply of food-fishes on the coast of New England and especially of Maine. The fact, as stated, needs no question; it is too patent to the experience of every man who has been interested in the fisheries, whether as a matter of business or as an amateur. An examination of the early records of the country in which the subject is referred to cannot fail to convince the most skeptical.

We are all very well aware that fifty or more years ago, the streams and rivers of New England emptying into the ocean were crowded, and almost blockaded at certain seasons, by the numbers of shad, salmon and alewives seeking to ascend, for the purpose of depositing their spawn, and that, even after these parent fish had returned to the ocean, their progeny swarmed to an almost inconceivable extent in the same localities, and later in the year descended to the sea in immense schools. It was during this period that the deep sea fisheries of the coast were also of great extent and value. Cod, haddock, halibut and the line fish generally, occupied the fishing grounds close to shore, and could be caught from small open boats, ample fares being readily taken within a short distance of the fishermen's abodes, without the necessity of resorting to distant seas. Now, however, the state of things is entirely different. The erection of impassable dams upon the waters of the New England States, and especially of the State of Maine, has prevented the upward course of the anadromous fishes referred to, and their numbers have dwindled away, until at present they are almost unknown in many otherwise most favorable localities.

The fact has been observed, too, that with the decrease of these fish there has been a corresponding dimunition in the numbers of cod and other deep sea species near our coast; but it was not until quite recently that the relationships between these two series of phenomena were appreciated as those of cause and effect. Halibut, it is believed, can be reduced in abundance by over-fishing with the hook and line, but experiences in Europe and American coincide in the confirmation of the opinion that none of the methods now in vogue for the capture of fish of the cod family (including the cod, haddock, pollock, hake, ling, etc.) can seriously affect their numbers. Fish, the females of which deposit from one to two million of eggs each year, are not easily exterminated unless they are interfered with during the spawning season, and as this takes place in the winter and in the open sea (the spawn floating near the surface of the water) there is no possibility of any human interference

219

with the process. Still, however, these fish have become comparatively scarce on our coast, so that our people are forced to resort to far distant regions to obtain the supply which formerly could be secured almost within sight of their homes.

It is now a well established fact that the movements of the fishes of the cod family are determined: first, by the search after suitable places for the deposit of their eggs; second, by their quest for food. Thus, the cod, as a summer fish, is comparatively little known on the coasts of northern Europe; but as winter approaches, the schools begin to make their appearances on the northwestern coast of Norway, especially around the Loffoden Island, arriving there finally in so great numbers that the fishermen are said to determine their presence by feeling the sounding lead strike the backs of fish.

Here they spend several months in the process of reproduction, the eggs being deposited in January, and the fishery being prosecuted at the same time. Twenty-five to thirty thousand men are employed in this business for several months; at the end of which the fish disappear, and the fishermen return to their alternate occupations as farmers and mechanics. The fish are supposed to move off in a body to the Grand Banks, which they reach in early summer, and where they fatten up and feed until it is time to return again to the northeast. It is believed that the great attraction to the cod on the Banks, consists in great part of the immense schools of herring and other wandering fish, that come from the region of Labrador and Newfoundland seas, and which they follow frequently close to the
shore, so that they are easily captured.

It is well known that the presence or absence of herring determines the abundance of hake and cod on the Grand Manan Fishing Banks, the fishes of the first mentioned family having a peculiar attraction to carnivorous fish of all kinds. It is, however, the anadromous fishes of the coast which bring the cod and other fishes of that family close in upon our shores. The sea herring is but little known, outside the region of the Bay of Fundy, excepting in September and October, when they visit the entire coast from Grand Manan to Scituate, for the purpose of depositing their spawn; this act depending upon their finding water sufficiently cold for their purposes, a condition which of course occurs later and later in the season, in going south.

In the early spring, the alewives formerly made their appearance on the coast, crowding along our shores and ascending the rivers in order to deposit their spawn, being followed later in the season by the shad and salmon. Returning when their eggs were laid, these fish spend the summer along the coast; and in the course of a few months were joined by their young, which formed immense schools in every direction, extending outward, in some instances, for many miles. It was in pursuit of these and other summer fish that the cod and other species referred to came in to the shores; but with the decrease of the former in number the attraction became less and less, and the deep sea fishes have now, we may say, almost disappeared along the coast.

It is therefore perfectly safe to assume that the improvement of the line fishing along the coast of Maine is closely connected with the increase in number of alewives, shad and salmon; and that whatever measures are taken to facilitate the restoration of these last mentioned fish, to their pristine abundance, will act, in an equal ratio, upon the first mentioned interest. The most important of the steps in question are the proper protection of these spring fish, and the giving to them every facility need for passing up the streams to their original spawning grounds; this to be done of course by the contruction of suitable fishways and ladders.

The real question at issue in regard to the construction of these fishways is, therefore, after all, not whether salmon shall come more plentiful, so that the sportsmen can capture them with the fly, or the man of means to be able to procure a coverted delicacy in large quantities and at moderate expense. This is simply an incident; the more important consideration is, really, whether the alewife and shad shall be made as abundant as before, and whether the cod or other equally desirable sea fish shall be brought back to our coast, so that one who may be so inclined, can readily capture several hundred weight in a day.

The value of the alewife is not fully appreciated in our country. It is in many respects superior to the sea herring as an article of food; is, if anything, more valuable for export: and can be captured with vastly less trouble, and under circumstances and at a season much more convenient for most persons engaged in the fisheries.

I have already extended this letter to an unreasonable length, and must therefore bring it to a close, with the assurances, however, that all the propositions I have thrown out can be amply substantiated.

221

Spencer F. Baird
U.S. Commissioner of Fish and Fisheries"

Source: Maine Fisheries Commissioners Report, 1872.

• LONG FIGHT FOR FISHWAYS, PRESUMPSCOT RIVER -- 1875.

[Note: This excerpt from the 1875 Report of the Maine Fisheries Commissioners provides one reason for the failure of the State of Maine's efforts to restore alewives, shad and salmon in the 19th century.]

"Five years ago, the Commissioners of Fisheries for Maine made the attempt to have fish-ways constructed over the dams on the Presumpscot river. Their efforts and the wishes of the people were defeated by the determined opposition of the mill owners. Since then, the amendment of the laws led the people to hope that their long-entertained desire, to have fish restored to their river, might be gratified; and in response to their importunities your Commissioners visited the Presumpscot river, viewed the dams and obstructions, and held meetings with and consulted the owners in relation to the proposed fish-ways. As a general rule, there was but little opposition expressed; all seemed willing to comply with the requirements of the law. At a further hearing, which was requested and held at the Falmouth hotel, the parties there present argued for more time, and desired that a year more should be granted them. Your Commissioners willingly assented to the request, if the parties seeking the continuance would bind themselves in good faith to build at the expiration of that time. Their reply was a prompt and energetic refusal. In due course we made surveys, furnished plans, and defined a time within which the structures should be built, all of which were duly served upon the respective parties. In the mean time, an organized opposition was determined upon, to oppose execution of the law. In order to gain time, and in conformity to their expressed determination, "to do nothing this autumn, but to go into the Legislature this winter," an appeal according to the provisions of Sect. 26, Chap. 40, was taken in ten cases, before the County Commissioners. Your Commissioners were duly summoned to appear at Portland, and after a long, vexatious, and fatiguing trial, occupying with its unavoidable adjournments a number of days, a decision was rendered on the third day of November, in their favor, of every point at issue, in every one of the cases, by a unanimous vote of the Board of County Commissioners. If the Legislature sees fit in its wisdom to grant to the appellants in these cases the same lenient extension of time as was granted to the owners of the Augusta dam, on the Kennebec, we think at the expiration of the coveted time they will be met by a similar exhibition of gratitude in a demand for an indefinite postponement."

• RESTORING ALEWIVES TO MAINE -- 1932.

Excerpt of Report of the Maine Commissioner of Sea and Shore Fisheries for the Biennium of 1932-1934.

"The alewives which migrate to the rivers to deposit their spawn are not protected as they should be and consequently bring only a small revenue into the state, whereas if properly protected and adequate fishways maintained these fish would increase in abundance by leaps and bounds. This fact has been demonstrated at Duck Trap Stream, a small stream which enters the ocean at Lincolnville. For a great many years two mills were maintained on this stream and tight dams prevented the alewives from ascending the river further than dams prevented the alewives from ascending the river further than the first dam, consequently their spawning ground could not be reached and the spawn cast was immediately destroyed and for many years alewives were not known in or near Lincolnville. A few years ago the mills were closed, the dams went out and now every spring finds a larger number of alewives ascending the stream to spawn. Protection is all that is necessary to increase and bring back to normal an almost depleted industry which once was a valuable asset to the state and provided hundreds of families with the comforts of life and was available as an abundant, natural food supply for the people of not only Maine and New England but of the United States. Why should not a sufficient amount be appropriated to protect this branch of the industry?"

• FISHWAYS FOR ALEWIVES -- 1967.

"At one time, over 3,000,000 pounds of shad were taken from Maine rivers, along with tons of alewives and salmon. In 1825, the St. Croix River was dammed near Calais; in 1830, the Penobscot River was dammed at Old Town; and in 1837, the Kennebec River was dammed at Augusta. Notwithstanding the provisions of the legislative charter authorizing the Augusta dam and calling for an adequate fishway, none was built; and this river and its tributaries, like the St. Croix and Penobscot, were closed to Atlantic salmon, shad and alewives. Other Maine rivers suffered a similar fate, and sea-run fish all but disappeared from the principal rivers. Only with the recent construction of fishways have Atlantic salmon and alewives begun making a comeback. Shad are seldom seen ....

"The fishway has been most successful in restoring a run of desirable fish when installed in a dam at the outlet of a lake or in the smaller, pollution-free coastal streams containing few dams. A fishway is desirable in a dam at a lake outlet when the lake provides good habitat for landlocked salmon and trout and when spawning and nursery areas are below the dam. Without a fishway, neither the adult fish nor their young could return to the lake. Where lakes are used by spawning alewives, these fish, of course, must have access to them. Sea-run fish have made an excellent comeback in coastal streams that have been opened to their migration, since it has been relatively easy to install fishways in the few low-head dams. The fish, fresh from the sea, are still strong, and there is little or no pollution to deter them."

Source: Decker, Laurence. 1967. <u>Fishways in Maine</u>. Maine Department of Inland Fisheries & Wildlife. Augusta, Maine.

• ATLANTIC COD EAT ALEWIVES -- 1972.

"Cod particularly go after the schooling fishes -- herring, menhaden, alewives -- and in the northern part of their range, around Newfoundland, for example, cod voraciously chase capelin (Mallotus) ... Each spring, hundreds of cod weighing up to 35 pounds are caught from the banks of the Cape Cod Canal in Massachusetts. The cod pursue the springtime schools of spawning alewives, a small herringlike fish, and are caught by anglers using alewives for bait."

Source: Jensen, Albert. 1972. <u>The Cod</u>. Thomas Y. Crowell. New York.

Scraping Topsoil for Profit or What Can You and I Do?

A friend recently bought 50 acres of land and a small ranch house in Rome, Maine at the headwaters of a spring brook called Rome Trout Brook, called that because it does, or at least used to, be the home of native brook trout. The land had been on the market for a few years at a reduced price, in part because the Town of Rome ran an industrial dump there in the 1950s and 1960s on top of the aquifer of the brook. So who knows what seeped in. The bank was not pleased about the potential for liability, hence closing on it was a bitch.

He got the land only because a previous buyer, who met the asking price, let on to the owner he intended to clearcut all the woods and bring in bulldozers to scrape all of the topsoil from the pastures to sell to landscaping contractors and re-open the old sand pits and gravel pits on the land and gouge them out too. Hearing that *precis*, the owner declined to sell.

Do we use our brief time on Earth to scrape away all the topsoil or to let it grow richer?

When my Dad and Tim and I each hit 30, it sunk into our misshapen native Massachusetts noggins how badly we had been robbed of our birthright: clean, unobstructed rivers filled with all of the native fish and wildlife which had lived in them since the thawing out of the

225

last Ice Age, when Paleo-Indians roamed the tundra of New England hunting caribou and wooly mammoth, fishing for arctic char, camping on George Banks, which was then dry, and making exquisitely fine fluted spear points from stones traded from Pennsylvania to Labrador. It also sunk into our noggins how much of it was still here, in our backyard in Easton, Massachusetts, and if certain little levers were pushed the right way, we could get much of it back.

But to get anything done, Tim and I had to figure out what we had lost. Our dad told us a lot. Joe Cardoza of Easton told us even more than Dad knew. But even they didn't know much. Even in their 60s and 70s they were too young, like us, born very late to the party. So Tim and I had to hit the bricks to the buildings with the books and dig up the words of those who saw what nobody alive has ever seen. These folks are long dead; their words, evidence, petitions and pleas silent in obscure archives and alcoves. For us, the mustiness of these catacombs only added urgency to the faded echoes of their screams. Like the narrator in Edgar Allen Poe's *Cask of Amontillado*, the proprietors of conventional wisdom long ago walled these voices behind layers of brick and mortar, deeming them insane and unreasonable. But as the poet Linton Kwesi Johnson said, "It's a room full of fact. You can't walk out." Or as we say paddling down the newly restored six mile reach of the lower Sebasticook River in Winslow, Maine, "It's a fack."

In the winter of 2004 I spent a week in the Maine Legislative and Law Library at the State Capitol building copying down every amendment made to every fish passage law enacted by the Maine and Massachusetts Legislatures (since Maine was part of Massachusetts until the Missouri Compromise of 1820).

Upon ingesting this brick of dead tree, it was apparent the laws requiring passage for fish at dams in New England were much stronger in the early 1700s than they are today; and as each decade went by, the laws became weaker and weaker, almost exactly in proportion to the increasing scarcity of the fish the laws were supposed to protect. The stronger the fish passage law, the less likely to be enforced. The weaker the law, the less it mattered if it was enforced. In 2010, standing chest-deep in the Kennebec River next to my house, once home to 100,000 Atlantic salmon as long as my leg, with a run of just four salmon that year, this brick of 'protective' laws on my kitchen table reminded me of confetti on a pauper's grave.

Today, Maine and Massachusetts have state fishway laws which apply to non-federal dams, ie. dams that are not used for hydroelectricity. Under these laws, the state's Commissioner of Fisheries & Wildlife *may* require a dam owner build fish passage at their own expense. Or may not.

State Fisheries Commissioners have almost never used this legal authority to require fishways at dams that destroy native fish populations. This is not because all of the small dams across New England are already equipped with effective fishways. Nearly all of them lack fish passage. The reason these laws are never used is because the states' Fisheries Commissioners, for reasons as inscrutable as Melville's Bartleby, simply prefer not to use them.

In 1968, the Maine Supreme Court ruled in *Dumont v. Speers* that citizens have no legal right to compel the state Fisheries Commissioner to even *consider* requiring fish passage at a dam, even if the dam is causing the entire Kennebec River to be dead in perpetuity. This ruling was used by the Maine Supreme Court in 2008 in *Friends of Merrymeeting Bay et al. v. Maine BEP*, which asked the Maine Board of Environmental Protection to stop the killing of thousands of pregnant, female American eels in the turbines of hydroelectric dams on Maine's Kennebec River. In 2008 the Court, citing *Dumont v. Speers*, ruled Maine citizens have no right to demand state environmental agencies use their legal authority under the U.S. Clean Water Act to stop the negative effects of privately owned dams on publicly owned rivers, including massive fish kills.

For this reason it is essential that state fish passage laws in New

England be restored to the language used in the 1700s, when the law required dam owners to not block migrating fish, to provide passage for fish, or to rip down their dam. Changing these laws is not easy, however. In the past four years, the Maine Legislature has twice rejected bills to make these fish passage laws work on behalf of citizens who live near rivers rendered dead by impassable dams. This fight must continue.

Hydroelectric dams are by far the most damaging things on New England's large rivers -- the Connecticut, Merrimack, Penobscot, Kennebec, Androscoggin, Saco and Presumpscot. These dams are controlled by the Federal Energy Regulatory Commission (FERC), created by the Federal Power Act, which Congress enacted in 1920. The Federal Power Act was written to encourage the construction of large dams, regardless of how they affect rivers, fish and people. FERC today continues to apply this outdated mandate. Through the FPA, Congress instructed FERC to issue licenses to new and existing dams with terms of 30-50 years, ostensibly so dam developers could amortize the building costs over many decades, thus helping them obtain the bank loans necessary to build the dam. But as Don Shields of Bangor said in 1994, all of New England's dams have long been paid for 100 times over. There is nothing left to amortize.

The Federal Power Act is the *Dred Scott* decision for America's rivers, the native animals who call them home, and innocent little kids who would like to grow up along their backyard rivers and have them not be dead. Because of the extremely long license terms issued to dams (30-50 years), citizens usually have only one chance in their lifetime to have a voice in how the dam in their backyard is operated. Unless you formally intervene in a dam relicensing when you are learning to read, you will not have another chance until your kids are having kids.

The U.S. Clean Water Act requires liquid waste dischargers to apply for a new discharge license every five years. This license term allows for any faults or deficiencies in the existing license to be quickly fixed, and gives citizens the right to immediately go to court if they believe the sewage plant is breaking the law. In contrast, at hydroelectric dams, the Federal Power Act requires citizens to wait 30-50 years before they even have the chance to point out that a dam needs a simple fishway.

Because of the Federal Power Act, New England's large rivers are now in worse shape than our smallest rivers, and are in far worse shape than they were in 150 years ago. A one sentence change in the Federal Power Act would place hydro dams on equal footing with wastewater treatment plants by reducing their license terms from 30-50 years to 5 years. Without this change, hydro dam owners will continue to keep New England's largest rivers dead and crippled in perpetuity.

Dams, not pollution, wipe out New England's alewife and other migratory fish runs, and squarely addressing the issue of dams is the only way to restore these fish. If the issue of dams is not addressed, the fish will not come back.

Alewives spend most of their lives in the ocean, not in rivers or ponds. What they do there, and exactly where, is little known, except they eat oceanic plankton and try to avoid getting eaten. But sadly we know that giant industrial fishing trawlers plying New England's coastal waters most likely catch, kill and throw back dead (as 'bycatch') most of the alewives in the ocean that our few healthy rivers send them, undoing in a few massive hauls everything we do to help alewives by letting them get over a dumb old mill dam. Since technology seems to always outstrip humans' ability to not use it disastrously, stopping these trawlers from wiping out the 'seed corn' we and the alewives try to plant every spring is essential.

As my friend Bill Townsend said, it's our job to give state and federal agencies the spine to do what they say they want to do and what the law tells them they must do. Another aphorism of Bill's is that "we are the water that slowly wears away the stone." Given how long it took mile high glaciers to make a few five foot wide potholes in the granite in the Sandy River in Phillips, Maine, that schedule is a bit too tardy for me, and certainly for a mother alewife fresh from the sea, hopelessly beating her head against the granite toe of a 200 year-old mill dam.

Like crows chasing an eagle, lots of folks seem to like to sit in meeting rooms and conference calls getting all giddy about talking about the 'concept' of doing something to help restore the fish of New England's rivers. But when the talk comes to doing something that will actually do something, somehow the agenda turns to folks checking their calendars and seeing when next month they can fit in another meeting to discuss that topic in full. Alewives and our rivers are not in such bad shape today because there have been too few meetings. If meetings *about* alewives were alewives, you could walk across the Penobscot on their backs.

As a general rule building fishways at dams is a bad way to restore sea-run fish. This is because fishways are fragile structures which do not work well compared to the same river without the dam. Just a few sticks or logs or garbage in a fishway, or a beaver building a dam in them, can make them fail. This means they must be cleaned and maintained every day. As winter ice seeps into cracks in the poured concrete, even the sturdiest fishways fall apart after a few decades and must be rebuilt from scratch. Because rebuilding a fishway costs a lot of money, this moment inevitably causes people in the future to debate whether they care enough about the fish to foot the bill. If they don't, the fish run is killed. Removing the dam in question relieves people of this perennial mulling by letting the fish go up river regardless of whether the townfolk want to spend $100,000 to build a new fishway or not, thus preserving the fish runs for little kids who might deeply care, but are not allowed to speak at town meeting.

American shad, the giant cousin of the alewife, will not use most fishways, due to their size and dislike of entering tight, confined places. A small, narrow fishway at a dam, even it is used readily by alewives, can be a death knell for American shad. They rarely use them, and as we have seen on the Androscoggin River's head of tide since 1980, shad often kill themselves in fishways by bashing their heads against the concrete walls.

History shows our society goes through cycles of being interested in protecting and restoring our native fish runs and then gradually or suddenly losing interest. So long as the existence of a fish run depends on people in the future making sure a clunky fishway at the dam actually works, these fish run the risk of being extirpated whenever a town committee or state agency loses interest in making sure the fishway still works.

The entire alewife run of the Nemasket River in Middleborough, Massachusetts, now totalling 1-2 million fish a year and probably the oldest surviving alewife run on Earth, was nearly snuffed out in the 1970s when the people of the town of Middleborough, stopped keeping maintained the falling down fishways on a couple old dams just a few feet high which have served no purpose for 150 years.

At Chesemuttock, along U.S. Route 44 in Middleborough there is a tiny dam on the Nemasket River several miles below Assawompsett Pond. It's the site of an ancient Indian fish weir. While the dam is still there, and serves no useful purpose, the fishway is a very wide inclined ramp of natural stair-steps made more than a century ago. It works exceedingly well, since it is not really a fishway. It's more like a re-constructed part of the river itself.

A mile up river, at Wareham Street, are two modern dams, the upper about six feet in height, the lower three feet in height. A very long, narrow and elaborate fish way guides some, but not all, fish around the lower dam. Many alewives jump over the lower dam and then are stranded and stuck below the upper dam. The cost of building and annually maintaining the fishway far outstrips the cost of pulling out the two dams, which serve no useful purpose. The second, lower dam was built only to keep alewives from getting stuck below the upper dam and to shunt them to the fishway, which would not be necessary if the tiny little dams were just removed.

Up the Taunton River from Nemasket, at the Town River, its real name Nunketetest, in downtown West Bridgewater, is a beautiful park called War Memorial Park. Here is the well preserved site of a 1700s era iron foundry, used to make cannon from local bog iron during the Revolution. For many years after World War II it was a junk yard but the town bought it and has cleaned it up and landscaped it. At the top of the park is a tiny dam, four feet high, with an old, cracked concrete fishway tortuously climbing around it.

Several years ago, when it became apparent the fishway was so decrepit it needed to be replaced, the townfolk of West Bridgewater were forced to debate removing this tiny dam, about the size of a small seawall, or building a new fishway. Not unexpectedly, caterwauls were heard as loud as those of the catamounts which used to haunt the Hockomock Swamp above the tiny dam. It was claimed if this tiny four foot high dam were removed, the entire 6,000 acre Hockomock Swamp would be drained and dry up, notwithstanding basic physics showing it could not, since the elevation of the Hockomock is some feet higher than the 'heighth' of the tiny old dam in downtown West Bridgewater.

As my brother noted during this imbroglio, the people of Easton, Raynham and West Bridgewater have been trying for 300 years to drain the Hockomock Swamp, and all's they had to do was build a four foot dam four miles downstream and take it down the next day.

On the Satucket River in East Bridgewater, a few miles down from the Nunketetest, is the Cotton Gin Dam, which has not done anything but block fish and fall down for 100 years. It keeps alewives from reaching Robbins and Monponsett Ponds. Each year they beat their heads by the hundreds against its toe trying to get past. Tim and I have tried for 10 years to get the Commonwealth of Massachusetts to use their legal authority to order the dam owner to either build a fishway or remove the dam. The state won't tell the dam owner to obey the law. This dam was equipped with a functioning fishway 50 years ago and its broken remains can still be seen. We can't even get the Massachusetts government to require a dam owner to keep a previously existing fishway maintained. This is why fishways are not the answer.

On the Town River in Bridgewater is a very old, 20 foot high concrete dam powering and doing nothing. A 200 foot long fishway, built by the Works Progress Administration (WPA) in the 1930s, tries to guide alewives around the dam. Every year, Bridgewater DPW crews have to come down to the river and put a wall of sandbags across the river channel to keep alewives from swimming to the foot of the dam and force them toward the tiny fishway entrance 200 feet below it. This one old dam is the sole reason why there are just a few thousand, rather than hundreds of thousands of alewives swimming up to Lake Nippinicket ('The Nip") in the Hockomock Swamp each spring. There is no reason for the dam to exist.

Above War Memorial Park in West Bridgewater, the Nunketetest is fed by several brooks below the Hockomock Swamp. One crosses Crescent Street in West Bridgewater, flowing from a place called West Meadows near the Brockton line. In the 1960s some agency of the government decided to build a giant earthen dam across West Meadows Brook and flood out a few hundred acres of its flood plain as a recreation area. Not knowing what they were doing, these folks also wrecked one of the few spring-fed native brook trout streams in the area, Brown Betty Brook, where my father caught native brook trout as a kid in the 1940s and 1950s. Today, each spring, a hundred or so intrepid alewives try mightily to swim past the tiny four foot dam at the outlet of West Meadows but cannot because it is built to be impassable. Even at West Meadows, a dam built by our own government, for our enjoyment, the government will not build a fishway for alewives at the dam. So they come back every year and beat their heads against the plywood flashboards fruitlessly.

The Weweantitt and Sippican Rivers form the capillaries of upper Buzzards Bay, the armpit of Cape Cod. The Weweantitt has the southernmost sea-run rainbow smelt run in North America, and lots of alewives and eels. There is a tiny old dam many yards below the river's head of tide, the Horseshoe Pond dam, built to power an iron foundry in the 1800s. The ground around the dam and the riverbed of the Weweantitt below it is made mostly of weird black, worm-shaped pieces

233

of iron slag skimmed from crucibles when McKinley was President.

In the mid 1990s, the U.S. Environmental Protection Agency allocated a gob of money to restore the health of Buzzards Bay and called it the Buzzards Bay National Estuary Project. A key goal of the Project was to identify and remove tidal barriers in the brooks, rivers and salt marshes that surround Buzzards Bay. The Weweantitt is the largest river flowing into Buzzards Bay and has an old impassable dam, the Horseshoe Pond dam, well below its head of tide.

Not long after the Buzzards Bay National Estuary Project was formed, it announced a plan to spend half a million dollars building a brand new dam at the site of the Horseshoe Pond Dam. My brother and I were confused. If the purpose of the Buzzards Bay National Estuary Project is to identify and remove tidal barriers, why weren't they trying to pull out the Horseshoe Pond dam, the largest tidal barrier in Buzzards Bay? Tim was told by the Project's executive director, Joe Costa, that he did not believe the dam was a tidal barrier because he believed the dam was well above the river's head of tide.

So Tim, who lived a couple miles up the road in South Middleborough, did some investigating. By standing at the foot of the Horseshoe Pond dam at high and low tide and watching the water go up and down over his legs and waist, he learned the Horseshoe Pond dam was below the head of tide.

But to be absolutely certain, Tim replicated the experiment by having his two young kids stand in the river at the foot of the dam as the tide came in to see if he had forgotten what a 'tide' was. But no, Danny and Hallie reported that when the tide started to come in, the water in the Weweantitt at the foot of the dam went from their ankles to over their heads.

Tim duly submitted this data to Joe Costa. Joe, who admitted he had never really spend much time at the dam site, said was unconvinced by Tim's double blind studies.

For the next several years, Tim and I and the kids went to the dam site and spent hours netting and hoisting hundreds of alewives and hundreds of thousands of baby eels ('glass eels') stuck below the dam over the dam. One day, Tim's son Danny, about 12 at the time, caught and brought in a 25 pound striper from just below the dam which had been chasing the alewives and eels trapped below the dam.

One May night, exploring an old concrete flume that shunted water from the dam to the iron works, I reached my hand in the water and pulled out a fish. It was an eight inch long rainbow smelt trying to get upstream. Tim later showed me photographs of tens of thousands of smelt eggs he saw strewn dead on the rocks below the dam during the spring smelt spawning time. Smelt eggs cannot survive in salt water. What he saw was the efforts of the last remnant of the Weweantitt smelt run trying desperately to get above the dam and cast their spawn, and being denied, casting them on the rocks where the saltwater came in and killed nearly all of them.

Late one summer night, on a full moon high tide, Tim and I went down with flashlights to the dam and watched as the tide crept over the concrete spillway and flooded it out. In the muddy shallows above the dam we saw hundreds of baby menhaden, a saltwater fish, just a half inch long, and tiny tan fish which look like a flounder, about two inches long, called 'hog chokers,' that never get much more than a few inches long. We submitted this evidence and photographs to Mr. Costa, who replied he was still certain the dam was far above the reach of the tide. Today, 10 years later, the dam is still there and fish still cannot get over it except at extreme high tides.

Despite these ongoing disasters, there are some good things happening in some unexpected places. They need to be noted and celebrated. There is a small stream, no wider than a one lane country road that enters the east side of the Penobscot River just above its head of tide.

It is called on the maps 'Blackman Stream' but has a much older Penobscot Indian name, Madamiscontis, 'fishing place for alewives.' Its mouth is on Route 178 in Bradley, Maine, on the opposite bank of the Penobscot River from Orono, Maine. Orono is named after a Penobscot Chief from the 1700s, Joseph Orono, and is home to the University of Maine which itself is on an island in the Penobscot River.

Madamiscontis drains three large ponds, the largest named Chemo, and was historically full of alewives until it was illegally dammed shut in the 1800s. It was a place I visited a lot as a student at Orono and a place I believed would never be brought back to life because it was too small and insignificant for the 'right people' to care about. In the 1800s a man named Benjamin Shaw had the poorly paid and unenviable position of being the 'fish warden' for Penobscot County, tasked with trying to make dam owners obey the law by keeping their dams passable for alewives, shad and salmon. In an 1848 memo to his boss, now in the Maine State Archives, Mr. Shaw wrote:

"I am impatient to have liberty to open Mr. Blackman's Stream in Bradley which should never have been closed; and bid welcome the finney Millions to return as of old to the great joy of themselves and the Inhabitants of the Region roundabout. I am informed that Elwives have been seen lingering round below that dam every year since the Law gave leave to close it against them. A good Fishway may be made cheap and be profitable to the owner and all the Country roundabout."

Mr. Shaw's great ambitions went unfulfilled for 152 years. But in 2010, after 10 years of work, volunteers from the Maine Council of the Atlantic Salmon Federation, led by wildlife scientist Ray "Bucky" Owen of Orono, finally got the funds, the partners, and the umpteen permissions and permits to get fish passage up Blackman Stream in Bradley and Eddington, Maine. In May 2010, the Maine Dept. of Marine Resources, led by Nate Gray, released 1,000 sea-run alewives from the Kennebec River into Chemo Pond, to make 'Mr. Blackman's Stream' once again mean Madamiscontis, place of alewives.

There is a Penobscot Indian creation story called Glooskap and the Frog. A malevolent giant man-frog, Oglebamu, decides to squat in the bed of the upper Penobscot River and keep all the water for himself and away from the villages and people below.

The story climaxes when Glooskap, 'the man from nothing,' as tall as the tallest white pine, dispatches the giant man-frog and lets the Penobscot flow freely again. Today, I see the face of the man-frog, Oglebamu, in the stone and concrete faces of all the hundreds of dams across the region, which like ghost lobster traps keep these rivers dead, kill the fish trying to pass them, and keep kids from enjoying their birthright of the natural wonder these rivers used to freely provide. So let us go forth and be our own Glooskap.

Last Song for the Salmon

The salmon, the leaper

Is now just about gone.

Their rapids that thundered

Are now still as ponds.

Concrete walls tall as prisons

Called dams keep them out.

Like mall parking lots

Where trees used to sprout.

And we sing our song for the salmon.

No one has an answer

No one has a plan.

We all know who did it

But we can't find the man.

He's hiding out somewhere

We choose not to see.

But he's always been standing

between you and me.

And we sing our song for the salmon.

I once waded a month

To put a hook in your mouth.

You had fought for an hour

When I pulled you out.

You died in my hands

And your eyes quietly closed.

What I felt I won't tell

What you thought no one knows.

And we sing our song for the salmon.

You were the last of your kind

And now I am too.

The last of my kind

To ever know you.

Like a dream disappears

Once you're awake

Like the branch that you bend

Is the one that you break.

And we sing our song for the salmon.

As the memories wash out

The ignorance flows.

In a flood down the river

To the ocean it goes.

If excuses were salmon

We'd have quite a few.

They'd be long as our legs

And wearing our shoes.

And we sing our song for the salmon.

If I was the river

And you were the fish.

I'd let you swim up and down me

And do as you wish.

Because without you

There's not much left of me

Just a long lonely ditch

Falling into the sea.

And we sing our song for the salmon.

We sing our song for the salmon.

So we sing our last song for the salmon.

-- Douglas H. Watts

God's Woods

Have you ever gone
for a walk in the fall
when God with bright colors
has painted it all.

Have you seen the sly deer
he's as smart as can be
I never see him
before he sees me.

Have you noticed the beech tree
with trunk smooth and grey
and its coppery leaves
that don't fall but they stay.

Have you seen the geese
overhead in a V
as they call to the world
"just let us be free."

Have you sat near a meadow
in the haze damp and grey
and watched while the sun
seems to melt it away.

Have you seen the chipmunk
hiding acorns in clumps
he resembles a child
with a case of the mumps.

Have you followed the squirrel
from tree to tree
and welcomed the return
of the chickadee.

Have you seen the great Oak

with leaves scarlet and gold
and the maples so varied
red, orange and bold.

Have you sat by a brook
as it babbles along
and the splash of a trout
interrupted its song.

If you've seen any of these
then you must surely know
that God only God
could put on such a show.

-- Allan E. Watts

References & Sources

[Note: Most of the primary historic documents in this volume were transcribed from the original copies at the Maine State Archives, State House Complex, Augusta, Maine. Excerpts of laws are from the official volumes at the Maine Legislative & Law Libary, Maine State House, Augusta, Maine. Secondary and additional reference sources are listed below.]

American Annual Register of the Years 1827-8-9. Vol. 3. Second Edition. 1835. New York, New York: William Jackson, and E. & G. W. Blunt.

Atkins, C.G., N. Foster. 1867. Report of Commission on Fisheries in: Twelfth Annual Report of the Maine Board of Agriculture, 1867. Stevens & Sayward, Printers to the State. Augusta, Maine.

Atkins C.G. and N. Foster. 1869. Maine Commissioners of Fisheries, First Report. Stevens & Sayward, Printers to the State. Augusta, Maine.

Baxter, James P., editor. 1910. Documentary History of the State of Maine Containing the Baxter Manuscripts. Vol. 22. Maine Historical Society. Lefavor-Tower Company. Portland, Maine.

Belding, David L. 1920. A Report on the Alewife Fisheries of Massachusetts. Mass. Dept. of Conservation, Division of Fisheries & Game. Boston, Mass.

Bigelow, H. B., and W. C. Schroeder. 1953. Fisheries Bulletin: Fishes of the Gulf of Maine. U. S. Fish and Wildlife Service, Washington, D.C.

Boardman, Samuel L. 1864. Aquaeculture: in Ninth Annual Report of the Secretary of the Maine Board of Agriculture. Stevens & Sayward, Printers to the State. Augusta, Maine.

Burt, Henry M. 1899. The First Century of the History of Springfield. The Official Records from 1636 to 1736. Vol. II. Published by the author. Springfield, Mass.

Champlain, Samuel de. The Works of Samuel de Champlain, Vol. I-IV with portfolio. Edited by H.P. Biggar. Toronto: Champlain Society, 1922-1936.

Collins, Michael. <u>Carrying the Fire: An Astronaut's Journeys</u>. New York: Farrar, Straus and Giroux, 1974.

Cook, Sanger Mills. 1966. <u>Pittsfield on the Sebasticook</u>. Furbush Roberts Printing Company. Bangor, Maine.

<u>Collections of the Maine Historical Society, Vol. 1</u>. 1831. Day, Fraser & Co. Portland, Maine.

<u>Collections of the Maine Historical Society, Second Series, Vol. 6</u>. Maine Historical Society. Portland, Maine.

Dole, Samuel T. 1974 reprint. <u>Windham in the Past</u>. Windham Historical Society. Windham, Maine.

Eckstorm, F.H. 1941. <u>Indian Place-Names of the Penobscot Valley and the Maine Coast</u>. University of Maine Studies, 2nd Series, no. 55. Orono, Maine.

<u>Falmouth Gazette and Weekly Advertiser</u>. 1785. Falmouth, Maine. Available on microfilm at the Maine State Library, Augusta, Maine.

Fisher, Carleton Edward. 1970. <u>History of Clinton, Maine</u>. Kennebec Journal Press. Augusta, Maine.

Flagg, L.N. 2007. <u>Historical and Current Distribution and Abundance of the Anadromous Alewife (Alosa pseudoharengus) in the St. Croix River</u>. Maine Atlantic Salmon Commission, Department of Marine Resources. Augusta, Maine.

Goode, G.B. 1887. <u>The Fisheries and Fishery Industries of the United States. Section V. History and Methods of the Fisheries</u>. U.S. Government Printing Office.

Gookin, Daniel. 1792. <u>Historical Collections of the Indians in New England</u>. Belknap and Hall, Apollo Press. Boston, Mass. 2000 Reprint Edition by Ayer Company Publishers. North Stratford, New Hampshire.

Goold, William. 1997 reprint. <u>Portland in the Past with Historical Notes of Old Falmouth</u>. Heritage Books. Bowie, Maryland.

Hanson, J.W. 1852. <u>History of Gardiner and Pittston</u>. William Palmer, Publisher. Gardiner, Maine.

Hardy, K. 2009. <u>Notes on a Lost Flute: A Field Guide to the Wabanaki</u>. Downeast Press. Camden, Maine.

Hoffman, K. 2008. The Maine Legislature's Bill: An Act to Stop the Alewives Restoration Program in the St. Croix River -- Have the Canadians and the Biologists Gone Berserk? Ocean and Coastal Law Journal, Vol. 13, No. 2. University of Maine School of Law. Portland, Maine.

Holmes, E., C.H. Hitchcock. 1862. Second Report Upon the Natural History and Geology of the State of Maine. Augusta, Maine.

Jensen, Albert. 1972. The Cod. Thomas Y. Crowell. New York.

Josellyn, John. 1674. Colonial Traveler. A Critical Edition of Two Travels to New England. Paul J. Lindholt, editor. University Press of New England. 1988.

Journals of the Massachusetts House of Representatives, 1776-1777. Available at the Maine Legislative and Law Library. Maine State House. Augusta, Maine.

Kennebec Journal, edition of May 26, 1829. Available on microfilm at Maine State Library. Augusta, Maine.

Kershaw, G.E. 1975. The Kennebeck Proprietors, 1749-1775. Maine Historical Society. Portland, Maine.

Kircheis, F.W., J.G Trial, D.P Boucher, B. Mower, Tom Squiers, Nate Gray, Matt O'Donnell, and J.S. Stahlnecker. 2004. Analysis of Impacts Related to the introduction of Anadromous Alewife into a Small Freshwater Lake in Central Maine, USA. Maine Inland Fisheries & Wildlife, Maine Department of Marine Resources, Maine Department of Environmental Protection.

Kircheis, F. W. 2004. Sea lamprey, *Petromyzon marinus*, L. 1758. F. W. Kircheis LLC, Carmel, Maine.

Lichatowich, J. 1999. Salmon Without Rivers: A History of the Pacific Salmon Crisis. Island Press. Washington, D.C.

Massachusetts Archives. Some documents pertinent to Maine available on microfilm at Maine State Library, Augusta, Maine.

Massachusetts Laws, Acts and Resolves. 1730-1820. Numerous volumes. Available at Maine Legislative & Law Library, Augusta, Maine.

McClellan, Hugh D. 1903. History of Gorham, Maine. Smith & Sale Publishers. Portland, Maine.

Monagle, W. 2007. <u>A Brief History of Lake Management in the Cobbossee Watershed District</u> *in* <u>Lakeline</u>. North American Lake Management Society. Madison, Wisconsin.

Nash, C.E. 1904. <u>The History of Augusta</u>. Charles E. Nash & Son. Augusta, Maine.

Natural Resources Defense Council. 2011. <u>Petition to the U.S. Secretary of Commerce List Alewife and Blueback Herring as Threatened Species and to Designate Critical Habitat</u>. NRDC. New York, NY.

Perley, M. 1852. <u>Reports on the sea and river fisheries of New Brunswick, 2nd edition</u>. J. Simpson, Fredericton, New Brunswick, Canada.

Petersen, J.B., 1991. <u>Archaeological Testing at the Sharrow Site: A Deeply Stratified Early to Late Holocene Cultural Sequence in Central Maine</u>. In: Occasional Publications in Maine Archaeology, vol. 8. Maine Historic Preservation Commission, Augusta.

Petersen, J.B., Robinson, B.S., Belknap, D.E., Stark, J., Kaplan, L.K., 1994. <u>An Archaic and Woodland Period fish weir complex in central Maine</u>. Archaeology of Eastern North America 22, 197–223.

Pierce, Joshua. 1862. <u>A History of the Town of Gorham, Maine</u>. Foster & Cushing and Bailey & Noyes. Portland, Maine.

Pory, John. 1622. Letter of John Pory to the Earl of Southhampton. In: <u>Three Visitors to Early Plymouth</u>. Reprinted by Plimoth Plantation. Plymouth, Mass.

Robinson, B.S. G. Jacobson, M.G. Yates, A.E. Spiess, E.R. Cowie. 2009. <u>Atlantic Salmon, Archaeology and Climate Change in New England</u>. J. Archaeol. Sci./doi:10.1016/j.jas.2009.06.001

Shelton, George. 1895. <u>History of Deerfield. Vol. 1</u>. Published by the author. Deerfield, Mass.

Spiess A. E., R. Lewis. 2001. <u>The Turner Farm Fauna: 5000 Years of Hunting and Fishing in Penobscot Bay, Maine</u>. Occasional Publications in Archaeology, vol. 11. Maine State Museum, Maine State Historical Preservation Commission, Maine Archaeological Society. Augusta, Maine.

Squiers, T. S., Jr. 1988. <u>Anadromous Fisheries of the Kennebec River Estuary</u>. Maine Dept. of Marine Resources. Augusta, Maine.

Stackpole, E.S. 1925. History of Winthrop, Maine. Merrill & Webber. Auburn, Maine.

The Centennial Celebration of the Settlement of Bangor, September 30, 1869. Published by Direction of the Committee of Arrangements. Benjamin A. Burr, Printer. Bangor, Maine.

Thoreau, Henry. A Week on the Concord and Merrimack Rivers, in: The Writings of Henry David Thoreau. Houghton Mifflin. Boston, Mass.

Thoreau, Henry. 1864. The Maine Woods. Ticknor and Fields. Boston, Mass.

Thurston, David. 1855. A Brief History of Winthrop from 1764 to October 1855. Brown Thurston, Steam Printer. Portland, Maine.

Trask, William, editor. 1901. Letters of Colonel Thomas Westbrook and others Relative to Indian Affairs in Maine, 1722-1726. George E. Littlefield. Boston, Mass.

Ulrich, L.T. 1990. A Midwife's Tale: The Life of Martha Ballard based on her diary, 1785-1812. Alfred A. Knopf, Inc. New York.

Watts, D.H. 2005. Historic Documents Related to the Anadromous Fisheries of the St. Croix River, Maine and Canada. Prepared for Maine Rivers. Hallowell, Maine.

Watts, D.H. 2004. Maine Historic Engineering Record. Dam No. 5. Cobbosseecontee Stream, Gardiner, Maine. Prepared for the Maine Historic Preservation Commission, Augusta, Maine: U.S. Army Corps of Engineers, Maine Field Office, Manchester, Maine.

Watts, D.H. 2002. Historic Records Related to the Anadromous Fisheries of the Presumpscot River and Sebago Lake, Maine. Prepared for Friends of Sebago Lake, Casco, Maine; Friends of the Presumpscot River, Windham, Maine. American Rivers, Washington, D.C.

Weston, Thomas. 1906. History of the Town of Middleborough, Massachusetts. Houghton and Mifflin. Boston, Massachusetts.

Williamson, W.D. 1832. The History of the State of Maine, Vol. 1. Glazier, Masters & Co. Hallowell, Maine.

Willis, T. 2009. How Policy, Politics, and Science Shaped a 25-Year Conflict over Alewife in the St. Croix River, New Brunswick–Maine. American Fisheries Society Symposium 69:000–000, 2009.

Willis, T. ; Bentzen, P. and I.G. Paterson. 2006. <u>Two Reports on Alewives in the St. Croix River</u>. Maine Rivers. Hallowell, Maine.

Willis, William. 1862. <u>History of Portland</u>. Maine Historical Society. Portland, Maine.

"From space there is no hint of ruggedness to it; smooth as a billiard ball, it seems delicately poised on its circular journey around the Sun, and above all it seems fragile. Is the sea water clean enough to pour over your head, or is there a glaze of oil on its surface? Is the riverbank a delight or an obscenity? The difference between a blue-and-white planet and a black-and-brown one is delicate indeed."

-- Michael Collins, Apollo 11 astronaut.

www.ingramcontent.com/pod-product-compliance
Lightning Source LLC
Chambersburg PA
CBHW060829170526
45158CB00001B/119